Developments in Geotechnical Engineering 6

DAMS, DAM FOUNDATIONS, AND RESERVOIR SITES

Developments in Geotechnical Engineering 6

DAMS, DAM FOUNDATIONS, AND RESERVOIR SITES

by

ERNEST E. WAHLSTROM

*Professor of Geological Sciences
Department of Geological Sciences,
University of Colorado, Boulder, Colo., U.S.A.*

ELSEVIER SCIENTIFIC PUBLISHING COMPANY
Amsterdam   Oxford   New York   1974

ELSEVIER SCIENTIFIC PUBLISHING COMPANY
335 Jan van Galenstraat
P.O. Box 211, Amsterdam, The Netherlands

AMERICAN ELSEVIER PUBLISHING COMPANY, INC.
52 Vanderbilt Avenue
New York, New York 10017

Library of Congress Card Number: 74–77587

ISBN 0 444-41236-0

With 165 illustrations and 23 tables.

Printed in The Netherlands

To Kay, with love and appreciation

PREFACE

The first dams built by early man were low earth or rock structures designed to impound and divert water for agricultural use. Today the face of the earth is dotted with small and large dams and reservoirs contributing in a variety of ways to the complex requirements of an expanding technologically advancing civilization. In this century remarkable strides have been made in utilizing natural earth materials to construct mammoth embankment dams, and concrete has replaced shaped stones in the building of low and high dams of a variety of geometric shapes.

In the total history of dam construction certainly many hundreds, and, if small embankment dams are included in the count, several thousand dams have failed. There are no accurate records of most of the failures and the extent of the damage that resulted from torrential inundation of lower areas as a consequence of their failure. Disasters owing to dam or reservoir failure for which careful documentation is available, indicate a variety of causes within the structures themselves or within the foundations.

In many parts of the world, especially in heavily populated, industrialized areas, many or most of the high quality dam and reservoir sites have already been utilized, and remaining sites, generally suffering from geologic defects in the foundations or lack of reservoir storage space, are requiring increasingly careful investigation by the most advanced methods of the geosciences and engineering prior to design and during construction so as to assure completion of storage or diversion facilities that pose only a minimum threat to those who live and work in the pathways of floods caused by failure.

Dams and reservoirs and the foundations on which they rest inevitably undergo changes with time. Some of these changes are slow and subtle and do not reveal their existence unless precisely and constantly monitored. Others, such as those caused by earthquakes, landslides, and unexpected floods are of short duration and usually can not be anticipated, in spite of the fact that the quality of knowledge concerning their prediction is constantly undergoing improvement.

The responsibility for the construction of a dam with maximum provisions for safety and the constant critical surveillance of the dam, reservoir, and foundation during its lifetime no longer belongs only to the engineer who builds the dam, but is shared by those who have special knowledge of hydrology, geophysics, geology, and rock and soil mechanics.

This book is written from the point of view of a geologist who has had a long and rewarding career working with engineers and fellow geoscientists in the design and construction of dams and tunnels. An attempt has been made to include between the covers of a small volume a review of (1) the theory and practice of geology and geophysics as they are applied in the investigation of proposed sites for dams and reservoirs, (2) procedures followed to ensure the continued safety of a dam during construction, and (3) geological processes and features of foundations that require continued evaluation after completion of a dam and filling of the reservoir behind it.

The writer is deeply indebted to his many associates who, through the years, have generously shared their knowledge with him. Particular thanks are due to engineering and administrative personnel of the Denver Board of Water Commissioners with whom the writer has worked closely in the planning and construction of an extensive system for collection and diversion of water for a large metropolitan area. A special debt of gratitude is owed to my fellow geologists, Professor L. A. Warner of the University of Colorado, and V. Q. Hornback of the Denver Board of Water Commissioners, who in close teamwork have shared many experiences and who have critically reviewed the manuscript.

Many of the photographs in the text were supplied by the Denver Board of Water Commissioners and by the U.S. Bureau of Reclamation, and credit is given in text titles for the courtesies extended by these agencies.

Finally, the manuscript for the book could not have been completed without the patient and highly valued assistance of my wife and co-worker, Kathryn K. Wahlstrom, who typed the manuscript and made many useful editorial suggestions.

CONTENTS

Chapter 1

ABOUT DAMS AND RESERVOIRS

INTRODUCTION

In many lands the face of the earth is dotted with lakes that occupy basins created by natural processes or reservoirs behind small to very large and technically complex dams built by man. The ultimate fate of all dams and reservoirs, unless they are carefully constructed and maintained, is deterioration and failure or filling by sedimentation. Every reservoir that impounds water behind a dam is a real or potential threat to those who live and work in flow channels below it, and, in some locations where earthquake shocks, movements along bedrock faults beneath dams, or collapse of large volumes of earth materials into reservoirs are distinct possibilities, even the most skilled design and continued maintenance may not preclude failures that are disastrous to life and property.

Modern technology utilizes the combined talents of engineers and geoscientists and has arrived at a state of development that enables a pronounced reduction of the risks that are inherent in dam and reservoir construction. This book emphasizes the geotechnical aspects of planning and construction of dams and reservoirs and reviews for the engineer and the geologist the origin, nature, and magnitude of problems in the foundation materials on which dams and reservoirs may be constructed.

PLANNING FOR DAMS AND RESERVOIRS

Prior to the turn of this century most dams were structures of low to moderate height and of simple design and purpose. With increasing demands for electric power, domestic, industrial, and irrigation water, and flood control, dams became larger and more complex and have been constructed in growing numbers. With the construction of large dams and reservoirs, many of them with multiple uses, engineering theory and technology have advanced with rapid strides, and safety in construction has become a primary consideration not only in new dams, but also in a continuing review of possible deterioration of older dams and reservoirs.

In densely populated areas, particularly in areas with advanced technologies, it would appear that all, or most of the high-quality dam and reservoir sites, already have been utilized and that the remaining sites are of inferior quality for one or several reasons. However, the need for storing additional water for irrigation or domestic use, for generation of electric power, or for flood control grows in intensity as the demands of an expanding population become progressively greater, and construction of dams and reservoirs, even in the face of increasing technological difficulties, probably çan not be forestalled. Redesign of existing collection and distribution systems, construction of pumped-storage facilities, and development of tidal storage dams and reservoirs offer attractive alternatives to acceptance of the situation as it presently exists.

In underdeveloped regions in many parts of the world vast areas remain where harnessing of the potential energy of small streams and large rivers and storing of waters in reservoirs is still possible and ultimately desirable. The technology that has been developed in areas of dense population and advanced culture can be applied in these areas for the good of all.

Engineers and geologists have by degrees arrived at a state of thinking where social responsibility is more important than the pride that goes hand-in-hand with construction of a major facility such as a large dam and reservoir. The emphasis now is not only on the need for, and the utility of, such construction, but also on the long-term safety of the project in terms of possible ultimate destruction of life and property.

Construction of dams and reservoirs in modern times requires the closest cooperation of engineers, soil-mechanics experts, and geologists in the planning, design, and construction of facilities that pose a constant threat to those who live and work below them, so as to assure a maximum degree of safety. In addition, existing dams and reservoirs require constant monitoring to determine whether changes with time are causing decay in the foundations, irreversible changes in reservoirs, or internal changes in dams that may result in disastrous failure.

KINDS OF DAMS

Dams range in size and complexity of construction from low earth embankments constructed to impound or divert water in small streams to massive earth or concrete structures built across major rivers to store water for irrigation, municipal use, hydroelectric power generation, or flood prevention. The kind of dam that is built and its size are a complex function of a demonstrated necessity for water storage or diversion, the amount of water available, topography, geology, and the kinds and amounts of local materials for construction. Although large embankment dams do not possess the graceful and architecturally attractive configurations of many concrete dams, they commonly require an equal amount of engineering skill in planning, design, and construction. The world's largest dams, as measured by the volumes of materials used in their construction, are embankment dams. Fort Peck Dam on the Missouri River in Montana, the largest of all existing embankment dams, required 125,000,000 cubic yards (92,000,000 m^3) of earth materials for its construction. In contrast, many of the world's highest dams are built of concrete, and many of them are 600 ft or more high.

Although there is no existing dam that exactly duplicates any other existing dam, or any dam that will ever be built, it is possible to identify several basic kinds of dams. Differences depend on their geometric configurations and the materials of which they are constructed. Under special circumstances features of the basic types are combined within a particular dam to meet unusual design requirements. A summary of basic types of dams is given in Table 1-1.

TABLE 1-1

Kinds of dams

1. Embankment dams.
    a. Homogenous dams, constructed entirely from a more or less uniform natural material.
    b. Zoned dams, containing materials of distinctly different properties in various portions of dams.
2. Concrete arch and dome dams.
    a. Single arch and dome dams.
    b. Multiple-arch and multiple-dome dams.
3. Concrete gravity and gravity-arch dams.
4. Concrete slab and buttress dams.
5. Dams combining two or more basic characteristics of the above basic types.

A brief characterization of the various basic kinds of dams and the natural conditions favoring their construction is given in following sections.

EMBANKMENT DAMS

A broad spectrum of natural and fabricated materials have been used in the construction of embankment dams. Controlling factors are the amounts and kinds of materials locally available for construction and the size and configuration of the dam. Many small embankment dams are built entirely of a single type of material such as stream alluvium, weathered bedrock, or glacial till. Larger embankment dams generally are zoned and constructed of a variety of materials, either extracted from different local sources, or prepared by mechanical or hydraulic separation of a source material into fractions with different properties. Where rock is used extensively, it may be obtained by separation from bowldery stream deposits, glacial till, slide-rock accumulations, or by quarrying.

Construction of an embankment dam requires prior investigation of foundation geology and an inventory and soil-mechanics study of materials available for

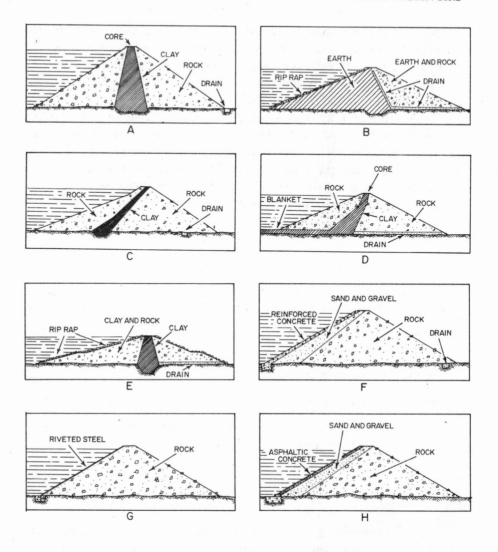

Fig.1-1. Idealized sections of several kinds of embankment dams. No scale.
A. Rock-fill dam with symmetrical clay core.
B. Earth and earth-rock dam.
C. Rock-fill dam with inclined clay core.
D. Rock and clay dam. Core is extended upstream as an impermeable blanket to reduce or eliminate seepage beneath dam.
E. Clay and rock dam with clay core.
F. Rock and gravel dam with reinforced concrete slab.
G. Rock dam with riveted steel facing.
H. Rock and gravel dam with slab of asphaltic concrete.

Fig.1-2. Construction photograph of Dillon Dam, Colorado. Note glory-hole spillway structure and control works in right foreground. (Photo courtesy of Denver Board of Water Commissioners.)

Fig.1-3. Palisades Dam, Idaho: an earth-fill embankment dam. (Photo courtesy of U.S. Bureau of Reclamation.)

Fig.1-4. Trinity Dam, California. An earth-filled embankment dam. (Photo courtesy of U.S. Bureau of Reclamation.)

emplacement in the embankment. An important element in a zoned dam is an impermeable blanket or core which usually consists of clayey materials, obtained locally. In the absence of such materials, the dam is built of quarried rock or unsorted pebbly or bowldery deposits, and the impermeable core is constructed of ordinary concrete or asphaltic concrete. Alternatively, in locations where natural impermeable materials are unavailable, embankment dams are built of rock or earth-rock aggregates and impermeable layers of reinforced concrete, asphaltic concrete, or riveted sheetsteel are placed on the upstream face of the dam.

Control of seepage through the dam or under it commonly requires installation of drains of porous materials within or immediately beneath the dam.

Embankment dams have been built on a great variety of foundations, ranging from weak, unconsolidated stream or glacial deposits to high-strength sedimentary rocks and crystalline igneous and metamorphic rocks. A particular advantage of an embankment dam, as compared with a concrete dam, is that the bearing-strength requirements of the foundation are much less. Minor settlement of an embankment dam owing to load stresses during and after construction generally is not a serious matter because of the ability of the embankment to adjust to small dislocations without failure.

Cross-sections of selected examples of embankment dams are shown in Fig.1-1. Photographs of embankment dams and appurtenant features are presented in Fig.1-2—1-4.

CONCRETE ARCH AND DOME DAMS

The ultimate complexity of design and analysis of stresses is attained in arch and dome dams. These dams are thin, curved structures commonly containing reinforcing, either steel rods or prestressed steel cables. Volume requirements for aggregate for manufacture of concrete are much less than in gravity and gravity-arch dams, but the competency of bedrock in foundations and abutments to sustain or resist loads must be of a high order. Arch dams usually are built in narrow, deep gorges in mountainous regions where access and availability of construction materials pose especially acute problems. At sites where abutments are not entirely satisfactory rock may be excavated and replaced with concrete to form artificial abutments.

Arch dams are of two kinds. *Constant-radius arch dams* commonly have a vertical upstream face with a constant radius of curvature. *Variable-radius dams* have upstream and downstream curves (extrados and intrados curves) of systematically decreasing radii with depth below the crest. When a dam is also doubly curved, that is, it is curved in both horizontal and vertical planes, it is sometimes called a "dome" dam. Curves that have been used in construction of arch or dome

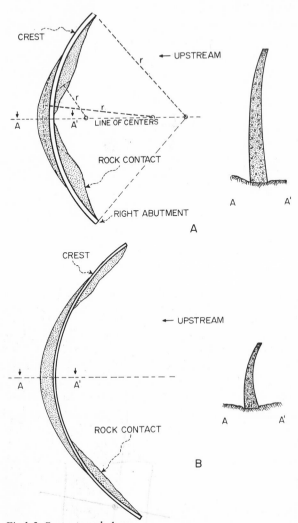

Fig. 1-5. Concrete arch dams.
A. Conventional arch dam with variable radius.
B. Arch dam, also called a "dome" dam. Arches are elliptical and of variable curvature, both vertically and horizontally. Configuration is that of Les Toules Dam, Switzerland.

dams are arcs or sectors of circles, ellipses, or parabolae. Some dams are constructed with two or several contiguous arches or domes, and are then described as *multiple-arch* or *multiple-dome* dams.

Engineering analysis of arch and dome dams assumes that two major kinds of deflections or dislocations affect the dam and its abutments. Pressure of water on the upstream face of the dam and, in some instances, uplift pressures from seepage beneath the dam, tend to rotate the dam about its base by cantilever action. In

← UPSTREAM

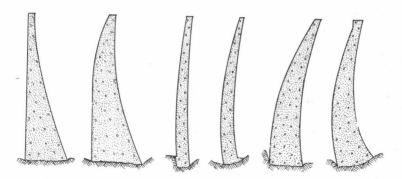

Fig. 1-6. Cross-sections of several varieties of arch dams. No scale.

Fig. 1-7. Anchor Dam, Wyoming. A thin-arch concrete dam. Overflow weir in notch in crest of dam discharges from a ski-jump type bucket near stream bed. (Photo courtesy of U.S. Bureau of Reclamation.)

Fig.1-8. Roosevelt Dam, Arizona. An arch dam built of shaped stones. Note layered sedimentary rocks in background. (Photo courtesy of U.S. Bureau of Reclamation.)

Fig.1-9. Monticello Dam, California. A concrete thin-arch dam. Note glory-hole spillway within reservoir. (Photo courtesy of U.S. Bureau of Reclamation.)

Fig.1-10. Williams Fork Dam, Colorado. This dam was built on an older, much lower dam. (Photo courtesy of Denver Board of Water Commissioners.)

Fig.1-11. Bartlett Dam, Arizona. A multiple-arch dam. Note spillway to left. (Photo courtesy of U.S. Bureau of Reclamation.)

addition, the pressure of reservoir water tends to flatten the arch and push it downstream, so that stresses are created which act horizontally within the dam toward the abutments. That portion of the bedrock abutment which receives the thrust from the load of reservoir water either by a tendency for downstream movement of the dam or flattening of the arch is called the *thrust block* and must be sufficiently strong to resist the forces acting on it without failure or appreciable dislocation. Simply stated, an arch dam utilizes the strength of an arch to resist the loads placed upon it by the familiar "arch action". It is clear that the foundation and abutments must be competent not only to support the dead weight of the dam on the foundation but also the forces that are directed into the abutments because of arch action in response to loads created by impounded reservoir water, and, in areas of cold climates, pressures exerted by ice forming on the reservoir surface. In regions of seismic activity consideration must also be given to the interaction of the dam and pulses of energy associated with earthquakes.

Two examples of arch dams in plan and section are illustrated in Fig.1-5. Several sections of arch dams are shown in Fig.1-6, and photographs of arch dams are presented in Fig.1-7—1-11.

CONCRETE GRAVITY AND GRAVITY-ARCH DAMS

A concrete gravity dam has a cross-section such that, with a flat bottom, the dam is free-standing; that is, the dam has a center of gravity low enough that the dam will not topple if unsupported at the abutments. Gravity dams require maximum amounts of concrete for their construction as compared with other kinds of concrete dams, and resist dislocation by the hydrostatic pressure of reservoir water by sheer weight. Properly constructed gravity dams with adequate foundations probably are among the safest of all dams and least susceptible to failure with time.

Final selection of the site for a gravity or gravity-arch dam is made only after comprehensive investigation of hydrologic, topographic, and, especially, subsurface geologic conditions. A favorable site usually is one in a constriction in a valley where sound bedrock is reasonably close to the surface both in the floor and abutments of the dam.

An important consideration in construction of a concrete gravity or gravity-arch dam is the availability, within a reasonable hauling distance, of adequate

Fig.1-12. Construction photograph of Gross Dam, Colorado. A gravity-arch dam with a centrally located spillway. (Photo courtesy of Denver Board of Water Commissioners.)

Fig.1-13. Folsom Dam, California. A concrete gravity dam. (Photo courtesy of U.S. Bureau of Reclamation.)

Fig.1-14. Canyon Ferry Dam, Montana. A concrete gravity dam. Note spillway structure upper right. (Photo courtesy of U.S. Bureau of Reclamation.)

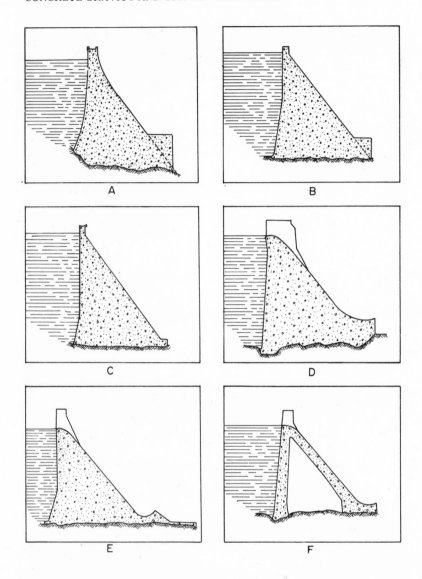

Fig.1-15. Typical sections of concrete gravity and gravity-arch dams.
A. Hoover Dam, Nevada-Arizona. Height about 725 ft (221 m).
B. Grand Coulee Dam, Washington State. Height 550 ft (168 m).
C. Fontana Dam, Tennessee. Height 450 ft (137 m).
D. Studen Kladenetz Dam, Bulgaria, Height 220 ft (67.5 m).
E. Sakuma Dam, Japan. Height 460 ft (140 m).
F. Topolintza Dam, Bulgaria, a hollow gravity dam. Height 278 ft (85 m).

Fig. 1-16. Hoover Dam, Arizona-Nevada. A massive concrete gravity-arch dam. (Photo courtesy of U.S. Bureau of Reclamation.)

deposits of aggregate suitable for manufacture of concrete, whether the aggregate is obtained from unconsolidated deposits or is quarried.

The simplest form of a gravity dam is one in which the top or crest is straight. Depending on the topographic configuration of a valley and the foundation geology, it may be possible to construct a gravity-arch dam which incorporates the advantages of mass weight and low center of gravity of a gravity dam with those inherent in an arch dam. In gravity-arch dams the requirements for sound rock in the abutments are somewhat more stringent than in simple gravity dams.

Some cross-sections of gravity and gravity-arch dams are shown in Fig. 1-15. Photographs of several gravity and gravity-arch dams are included in Fig. 1-12—1-14 and Fig. 1-16.

CONCRETE SLAB AND BUTTRESS DAMS

In locations where aggregate for concrete or for earthfill is in limited supply and the foundation rocks are moderately to highly competent, buttress dams provide a possible alternative to other kinds of dams. In cross-section buttress dams resemble gravity dams, but with flatter upstream slopes. In a buttress dam a slab of reinforced concrete upstream rests on a succession of upright buttresses which have

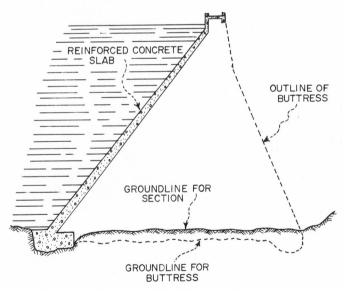

Fig.1-17. Section of a concrete slab and buttress dam.

Fig.1-18. Stony Gorge Dam, California. A concrete slab and buttress dam. Note water discharging in spillway. (Photo courtesy U.S. Bureau of Reclamation.)

thicknesses and a spacing sufficient to support the concrete slab and the load of the water in the reservoir exert on the slab.

A cross-section of a typical buttress dam is shown in Fig.1-17. A photograph of a buttress dam of intermediate height is shown in Fig.1-18.

APPURTENANT FEATURES OF DAMS

In previous sections of this chapter the characteristics and configurations of basic types of dams have been outlined, but no consideration was given to the various appurtenant features that enable use of a dam and the reservoir behind it for their intended purposes. In this section mention is made of various features that are incorporated into the designs of dams for control of flow of water impounded in the reservoir through or outside of a dam. Design of many of these features requires intensive prior investigations of hydrology, topography, and subsurface geology of the dam site. Following is a tabulation and description of several kinds of appurtenant features included in the construction of dams. Additional terms associated with dams are defined in a Glossary at the end of the chapter.

Coffer dams

Coffer dams usually are temporary structures built upstream from a dam to divert stream flow around the excavation for a dam. In valleys of steep profile diversion commonly is accomplished by a tunnel or tunnels in the walls of the valley. Commonly the diversion tunnels are put to further use to control flow from the reservoir either for drainage of the reservoir or for flow under pressure into a hydroelectric generating plant. In valleys of low profile diversion is by tunnels, canals, or by conduits which subsequently are buried by the dam. It is not unusual in embankment dams to incorporate the coffer dam into the larger embankment structure comprising the designed dam.

Fish ladders

Dams constructed on streams that are the migration paths for spawning fish commonly make provisions for movement of the fish up or in the vicinity of the downstream face of the dam. The facility that permits fish migration is usually called a "fish ladder".

Gates

Gates are devices installed in the tops of spillways to control the flow of water over the spillway.

Hydroelectric power plants

Many dams are constructed to generate hydroelectric power. The powerhouse is located at, or in the vicinity of, the toe of a dam or at some distance downstream. Flow of water into the power house is controlled by valves upstream from the dam, within the dam downstream, or in valve vaults excavated in rock outside of the dam.

Locks

Locks are movable dams or portions of dams utilized in navigation along rivers and canals.

Penstocks

A penstock is a sluice or conduit used for control of water flow, especially into a hydroelectric power plant.

Spillways

A spillway is designed to contain and control overflow of reservoir water when the reservoir is full. Spillways are, or should be, designed to accommodate flows during maximum flood stage so as to prevent damage to the dam and appurtenant features. Their size and location with respect to the dam is determined by the size and kind of dam, local topography, geology, and a careful review of the history of stream flow at the site of the dam.

Overflow of embankment dams outside of a spillway can have especially disastrous consequences so that safety usually requires a spillway capable of containing at least a hundred-year flood.

Spillways are located within or on the downstream face of a dam, outside of the dam on one side or the other, or within the reservoir, where water spills into a "glory hole" and passes through a shaft and tunnel or tunnels in the abutment of the dam. It is instructive to review the diagrams and photographs in Fig. 1-2–1-18 to observe some of the various locations of spillways with respect to dam structures.

Tunnels

Tunnels in bedrock outside of dams serve a variety of purposes. Flow through them is controlled by valves external to the dam or in valve chambers or vaults within the dam or in bedrock outside of the dam. Tunnels for control of the water level in the reservoir are commonly called *gravity tunnels* and serve a principal function in diverting water to some point downstream from the dam. Tunnels that transmit water under pressure to elevate the water to a higher level than the intake of the tunnel or to generate hydroelectric power are called *pressure tunnels* and usually require considerable competency in the rock through which they are constructed.

Valves and valve vaults

Valves control the flow of water through tunnels and penstocks. In many large dams the valves are installed in underground vaults or chambers to which access is gained downstream from the dam.

KINDS AND USES OF RESERVOIRS

There are several types of reservoirs as defined by their locations. The commonest type is a reservoir behind a dam in a valley. Increasing use is being made of tidal-storage reservoirs for power generation along coast lines and excavated reservoirs for water storage for municipal or other use. In the latter type the excavated material commmonly is employed tc construct an embankment on one or all sides of the reservoir.

Reservoirs and their associated dams serve many purposes including electric power generation, storage and diversion of irrigation water, storage of industrial and municipal water supplies, recreation, and flood control. Less frequent uses of reservoirs include storage and control of stream water for navigation, and storage of sewage and waste products from mining or manufacturing operations. In some instances fuel powered and thermonuclear power plants require large volumes of cooling water, and reservoirs are constructed for this purpose.

With increasing energy demands efforts are being made in many countries to utilize hydroelectric facilities to the maximum, and pumped-storage reservoirs are being built to increase useful electric power output. This trend makes it possible to review potential dam and reservoir sites which in the past were not constructed because of inadequate water supply or for other reasons.

Historically electric power generation using the force of water under a head, especially water stored in reservoirs, has been subject to the vagaries of nature, and only a few hydroelectric plants, mostly those along major rivers, have been sources of a steady, year-around supply of electricity. In many localities power from water stored in reservoirs is generated only intermittently during periods of peak load or greatest demand to supplement power from other steady sources.

There are times in electric-power consumption cycles when the total generating capacity of a system usually exceeds the demand, and the unused energy can be converted at low cost to potential energy for use during intervals of peak load. The potential energy in question is, of course, the energy associated with the head of stored water in a facility at a higher elevation than a hydroelectric generating plant. Reservoirs into which water is pumped for storage either for subsequent use in power generation or for diversion for other purposes are called *pumped-storage reservoirs*.

Two systems for utilizing pumped-storage reservoirs for generating hydroelectric power are schematically shown in Fig.1-19. In Fig.1-19A an upper reservoir is supplied by runoff and pumping. Power for pumping from the lower reservoir is produced locally by fuel or by electricity from a distant source. During periods of peak demand the potential energy of the water in the upper reservoir is converted to electrical energy at the power and pump station. Theoretically this system would work even in the absence of a runoff supply by repeated transfer of water from one reservoir to the other.

Fig.1-19B illustrates a more complicated system in which generators at lower elevations supply power for pumping into a central reservoir from storage basins at intermediate elevations. Part of the power in the system is generated by seasonal runoff water stored within and above the major storage reservoir.

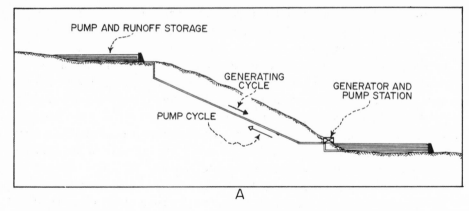

A

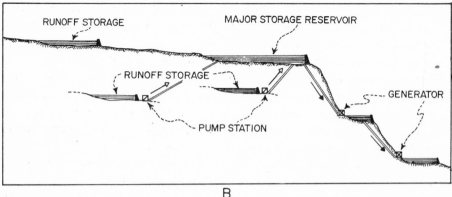

B

Fig.1-19. Schematic pumped-storage systems for generation of hydroelectric power.
A. Simple system using locally generated or an outside source of power for pumping.
B. Pumped-storage with power for pumping generated locally. Water for pumped-storage is
obtained from subsidiary reservoirs at intermediate elevations.

It is clear from an examination of Fig.1-19 that a great many possible
permutations in the details of design and construction depend on local conditions
where it is proposed to establish such a system.

SAFETY OF DAMS AND RESERVOIRS

There are no accurate records of the number of dams that have failed
throughout the history of their construction. However, ruins of dams built through
a long span of history in both ancient and more recent times indicates that the
number must be high, probably in the thousands, if dams of all heights are included
in the count. The magnitude of floods generated by dam failure or by collapse of
the walls of a reservoir are not necessarily related to the height of a dam. More

pertinent is the volume of water stored in the reservoir behind the dam, and the configuration of the valley below it, whether the dam is low or high.

The cost in human life, goods, and property damage of a flood generated by breaching of a dam or collapse of reservoir walls depends to a large extent on the magnitude of the flood and what lies in the pathway of the flood. With an expanding population in many parts of the world and an increasing occupation of floodplains by dwellings, commercial and industrial facilities, and highways, dams built long ago and dams built in recent times present a growing potential for massive destruction of life and property.

The causes of floods associated with failure of dams and reservoirs are numerous. In embankment dams a common cause of breaching is overtopping of existing spillways or waterlevel control facilities, although some failures have been attributed to slope failure, foundation subsidence, or earthquake damage. Failure of concrete dams usually is attributable to imperfect design or construction, to use of inferior materials in the dam, or to failure of foundation and/or abutment rocks. The disastrous flood from the Vaiont reservoir in Switzerland in 1963 was caused by the sudden collapse of Mount Toc into the reservoir, and indicates the need for careful geological examination of reservoir basins as well as dam sites prior to, during, and after construction.

Construction of a dam and reservoir imposes new loads on foundation materials. Initial adjustments in the dam and foundation occur as the dam is being built and as the load on the foundation is increased to a final load equal to the weight of the dam. Filling of the reservoir imposes additional loads not only on the floor and walls of the reservoir but also on the upstream face of the dam. As the reservoir level is decreased and increased these loads fluctuate, and a cyclic dynamic system of changing loads is superimposed on the static load of the dam on its foundation. Seepage of water through or beneath the dam may produce slow deterioration that may promote eventual failure.

Engineers have developed the theory of static and dynamic stress and strain relationships in dams to a high state of sophistication and, given a uniform foundation of competent materials, can predict the nature of the interaction between the dam and the natural materials below it and at its sides so as to allow for a large margin of safety. However, rocks and unconsolidated deposits on which dams are constructed almost never are uniform, monolithic bodies. Subtle to gross variations in composition and strength of original materials, together with changes that have occurred in them with geologic time such as fracturing and chemical alteration, have rendered most natural earth materials moderately to highly aniso-tropic. The geometry of the anisotropism, in some instances is so complex that mathematical stress—strain analysis is not possible, and resort must be made to personal judgments based on experience and empiricism.

The responsibility for the safety of dams and reservoirs no longer belongs only to the designer and builder but must be shared by those who have knowledge and understanding, however imperfect they may be, of the expected behavior of natural materials under the conditions superimposed by the loads of dams and reservoirs. Modern technology and social responsibility require that safe construction and maintenance of dams and reservoirs shall be the shared responsibility of engineers, geologists, and rock mechanics experts.

Recognition of the need for world-wide surveillance of dams and reservoirs with emphasis on their safety has resulted in the formation of the International Commission on Large Dams (ICOLD), a unit of the World Power Conference. Within individual countries increasing efforts are being made to regulate and maintain continued safety of dams and reservoirs through the close cooperation of engineering organizations and governmental agencies. In spite of these good works, dams continue to fail, and intensification of efforts to assure the safety of existing dams and reservoirs and those that will be built in the future is an increasingly urgent necessity.

GLOSSARY

This glossary includes only terms or expressions widely used in design and construction of dams, reservoirs, and appurtenant features and is not intended to be comprehensive. Specifically omitted are definitions of the detailed elements of construction of hydroelectric power generating units.

Abutment. That portion of the foundation, especially in the sides of a valley, which is in contact with a dam. Also, that portion of a dam which makes contact with and abuts against the foundation at the sides of a valley.

Active basin. The portion of a reservoir basin above a given elevation which can be used for power generation or other beneficial purpose.

Active pool. The same as *active basin,* which see.

Active storage. The same as *active basin*, which see.

Aggregate. Natural material used in the manufacture of concrete. Also any natural material, sorted or unsorted, used in dam or other construction. Aggregate for concrete commonly is obtained from alluvial stream deposits or from rock quarries.

Appurtenant feature. Any physical feature other than the dam itself which contributes to the operation of the dam and reservoir for its intended purpose or purposes.

Artificial abutment. An abutment, usually constructed from concrete, to sustain the lateral thrusts of an arch dam. Such abutments are constructed where existing topographic or bedrock geologic conditions are not adequate for the design of the dam.

Asphaltic concrete. An impervious mixture of aggregate and bitumens used in cores or upstream surfaces of embankment dams.

Axis of dam. A reference line used for control of surveying during construction of a dam. Commonly the axis defines the location of the upstream portion of the crest of a dam, whether the crest is straight or curved.

Bedrock. The natural, more or less undisturbed rock in the foundation of a dam.

Blanket. A thin horizontal or inclined layer of material forming a part of an embankment dam.

Blanket grouting. Shallow, systematic grouting with cement-water mixtures or chemical solutions of bedrock exposed in an excavation for a dam.

Block. Many concrete dams are built in sections or blocks. A section of a concrete dam emplaced within forms or contained between upstream and downstream forms and adjacent sections of the dam is a *block.*

Borrow area. The source area for natural materials used in dam construction.

Bucket. The curved bottom portion of a spillway. The bucket deflects upward and outward the water flowing down the inclined surface of the spillway.

Bulkhead. A structure built to resist rock pressure or to shut off water flow, as in a tunnel.

Buttress. A thin, erect, tabular concrete supporting member used in construction of slab and buttress dams. Also, a projecting structure providing lateral support to a rock face or a portion of a dam.

Cableway. A steel cable used in placing concrete in a dam and to transport excavated materials and construction materials over and above a dam.

Chute. An inclined open trough or lined canal through which water is discharged.

Cofferdam. A temporary dam designed to contain and divert water away from the excavation for a dam or other facility during construction. In some embankment dams the cofferdam is subsequently incorporated into the main, larger structure.

Clay blanket. A thin layer of impervious clay placed upstream from an embankment dam to reduce or eliminate seepage beneath the dam.

Compacted fill. Material in an embankment dam that has been compressed by rolling or impact vibration.

Concrete piling. Pillars of concrete driven vertically downward into unconsolidated materials below an embankment dam to reduce or eliminate seepage beneath the dam.

Control tower. A tower commonly constructed a short distance upstream from a dam and within the reservoir to control flow of water from the reservoir into conduits or tunnels.

Control works. Facilities such as valves and gates designed to control flow from the reservoir through, over, or around a dam.

Construction joint. A joint between adjacent blocks of concrete. Also, a joint,

usually nearly horizontal, between a layer of concrete and the next one
placed over it during construction.

Core. The central portion or zone of an embankment dam consisting of impervious
material.

Core trench. The trench excavated below the general level of the base of an
embankment dam and filled with the impervious material used to construct
the core.

Crest. The top of a dam.

Curtain. A zone of foundation grouting or piling parallel to a dam axis designed to
prevent or diminish seepage beneath the dam.

Curtain grouting. Grouting of foundation materials to produce a barrier to seepage
beneath a dam.

Cut-off. A fabricated structure or a grout-curtain placed to intercept seepage flow
beneath a dam.

Dam. A barrier, either natural or artificially constructed, that impounds or diverts
the flow of water, especially in a water course. Also, the body of water
confined by a dam.

Dead storage. Water in the lower elevations of a reservoir that is unavailable for use
or diversion.

Dike. A small embankment dam.

Drain. A facility for collecting and diverting water that seeps through a dam or
through the foundation of a dam.

Drainage holes. Drilled holes designed to intercept seepage water within or beneath
a dam.

Drainage prism. A geometrically shaped zone of permeable materials installed in or
below an embankment dam to intercept seepage.

Drawdown. Reduction of the water level of a reservoir.

Dumped fill. Material that is placed in an embankment dam without special
additional treatment, such as rolling.

Earth fill. Material consisting of earth excavated from a nearby borrow area used in
the construction of an embankment dam. The term is imprecisely defined but
is generally applied to materials containing abundant soil and clayey substances with or without rocky components.

Embankment. A raised structure built up from unconsolidated materials.

Fill. The natural material used to construct an embankment dam.

Filter zone. A porous zone in or below a dam designed to intercept and divert
seepage water.

Fish ladder. A structure built at the side or up the face of a dam to enable
migration of fish upstream and downstream.

Flash board. A wood plank or a steel member placed at the top of a spillway to
increase the storage capacity of a reservoir.

Foundation. The surface and the natural material beneath it on which a dam and appurtenant features rest.

Foundation cut-off. An excavated trench beneath or adjacent to a dam filled with impermeable material or a grout curtain designed to prevent seepage in the foundation beneath a dam.

Freeboard. That portion of a dam above maximum water level in a reservoir.

Gallery. A long, narrow passage inside of a dam used for inspection, grouting, or drilling of drain holes.

Gate. A movable facility for controlling flow of water over a dam through a spillway.

Gravity tunnel. A tunnel through which water flows without restraint under the force of gravity.

Grout. A mixture of water and cement or a chemical solution that is forced by pumping into foundation rocks or joints in a dam to prevent seepage and to increase strength.

Grout blanket. A grouted zone in the shallow portion of a foundation which has been treated to improve its strength and reduce its permeability.

Grout cap. A cap, usually consisting of concrete, through which grouting operations of foundations are performed.

Grout curtain. A zone in bedrock beneath a dam and parallel to its length that has been injected with grout to stop or reduce seepage beneath a dam.

Grout trench. A trench excavated to enable construction of a grout cap.

Grout veil. The same as a grout curtain, which see.

Grouting. The operation whereby grout is injected under pressure into openings in a dam or in its foundation.

Gut. A term sometimes used for the cableway above a dam used for transportation of construction materials.

Head. The hydrostatic pressure generated by water in a reservoir.

Headrace. The flow of water in the direction of a controlling valve or, more especially, through a conduit or tunnel toward a power-generating unit.

Headrace conduit. A conduit that conducts water under a head to a valve or into a power-generating unit.

Headrace tunnel. A pressure tunnel which conducts water from the reservoir to control works and ultimately into a power-generating unit.

Heel. The upstream contact of a dam with its foundation.

Hydraulic fill. Fill pumped or directed by channel flow into an embankment dam during construction.

Impervious blanket (or membrane). A thin layer of impervious material placed within an embankment dam or on the floor of a valley upstream from the dam to reduce or eliminate seepage through or beneath the dam.

Impervious core. A core in a zoned embankment dam consisting of impervious material.

Impervious material. Material, usually rich in clay and/or silt-size particles that resists penetration by water.

Inactive basin. That portion of the bottom of a reservoir that contains water that can not be put to beneficial use or drained from the reservoir.

Inactive storage. The storage in an inactive basin, which see.

Inspection gallery. A gallery within a dam which enables examination of the performance of the dam with time and reservoir filling.

Instrumentation. Devices installed on and within a dam to monitor cyclic or progressive changes during and after construction of the dam.

Intake. The entrance to any water transporting facility such as a conduit or a tunnel.

Intake structure. The structure built at the intake.

Left abutment. That portion of the dam that makes contact with its foundation on the left side of a valley as viewed from upstream.

Muck. A common expression for unconsolidated, usually wet and muddy natural material.

Natural abutment. An abutment in natural foundation materials. Contrasted with an *artificial abutment* which is constructed from concrete at the site.

Observation well. An excavated well or vertical borehole used to observe changes in seepage flow through or beneath dams during filling and drawdown.

Outlet. Any facility, such as the exit of a tunnel, from which water issues by controlled flow.

Outlet structure. A fabricated structure at the outlet of a canal, conduit, or tunnel.

Overflow section. That portion of a dam, usually occupied by a spillway, over which water above the spillway elevation flows.

Parapet. Usually construed to be a low protective wall along the crest of a dam.

Pendulum shaft. A narrow vertical opening in a dam used for surveying control during construction and subsequently for observation of deflections of the dam under load.

Penstock. A conduit, commonly steel pipe, leading from the reservoir to a power-generating plant downstream from the dam.

Pervious material. Material through which water flows with relative ease. Contrasted with *impervious material.*

Piling. Elongate, post-like steel or concrete members or steel sheets driven into a dam foundation to reduce or eliminate seepage.

Power intake. The intake to a conduit or tunnel which leads to a power-generating unit.

Power plant. The facility constructed at or near the downstream face of a dam to generate hydroelectric power.

Pressure tunnel. A tunnel which transmits water under moderate to high pressure.

Prestressing. Strengths of rocks in foundation and elements within concrete dams are increased by installation of steel rods or steel cables which are subjected to tensioning. The procedure that is followed is called *prestressing.*

Purge tunnel. A tunnel that is used to flush more frequently used tunnels of obstructions or deposits of sediment.

Relief well. An excavated well below a dam to collect seepage water in the foundation.

Reservoir. In the present context a reservoir is a basin, usually artificially created, that impounds and stores water.

Right abutment. The abutment to the right as observed from a point upstream from a dam.

Rock blanket. A layer of rocks placed on the face of a dam to prevent wave erosion of deeper materials.

Rock bolt. A threaded steel rod placed in a drilled hole and tensioned to increase strength of rock masses.

Rock fill. Rock aggregate placed in an embankment dam.

Rolled fill. Fill, usually rich in clayey or silty components, that is compacted by rolling, especially with "sheep's foot" rollers or vibratory compactors.

Saddle dike. A small dam built in a topographic low in the periphery of a reservoir basin.

Sheet piling. Plates of steel driven into the foundation of a dam to reduce or eliminate seepage below the dam.

Sluiced fill. Fill, usually clayey, placed in an embankment dam by running water.

Spillway. The structure on or at the side of a dam that contains and guides the flow of excess water supplied to a reservoir. Spillways inside the reservoir are called "glory holes" and consist of a vertical shaft and a tunnel which exits below the dam.

Spoil area. An area used to dispose of materials that are unwanted or surplus in dam construction.

Stilling basin. A basin downstream from a dam that receives the discharge from tunnels or conduits or overflow from a spillway.

Structural damage. Damage resulting from failure of a dam or its appurtenant features.

Surge tank or shaft. A vertical shaft above a pressure tunnel that provides equal pressures at the tunnel level in response to sudden pressure changes caused by increasing or decreasing the flow of water.

Tail water. The water issuing downstream from tunnels, conduits, or spillways.

Tail race. The movement of water below a valve or after it has passed through a power-generating plant.

Thrust block. That part of the foundation of an arch dam against which horizontal thrust is exerted by the dam as the reservoir behind it is filled.

Toe. The downstream contact of a dam with its foundation.

Tower. A vertical structure upstream from a dam designed to control flow of reservoir water through the dam into power-generating facilities.

Trash rack. The screening facility built at the intake end of conduits or tunnels to prevent entrance of debris.

Valve chamber. A chamber within a dam containing valves to control the flow of water from a reservoir.

Valve vault. An opening excavated in bedrock at the side of a dam and containing valves for control of flow from the reservoir.

Water stop. A membrane placed in joints in concrete dams to prevent seepage of water.

Weir. A channel of known cross-section which enables measurement of the volume of flow of water after calibration. The top of a spillway set into a concrete dam is also sometimes designated as a weir.

Zoned dam. An embankment dam in which materials of different properties are placed systematically in various portions of the dam.

REFERENCES

Hanna, F. W. and Kennedy, R. C., 1938. *The Design of Dams.* McGraw-Hill, New York, N.Y., 2nd ed., 478pp.

Johnson, W. E., (Chairman), 1967. *Joint ASCE—USCOLD Committee on Current United States Practice in the Design and Construction of Arch Dams, Embankment Dams, and Concrete Gravity Dams.* Am. Soc. Civil Engrs., 131pp.

Ono, Motoki (Publisher), 1967. *World Dams Today.* The Japan Dam Association, Tokyo, Japan, 436pp.

Rescher, O. J., 1965. *Talsperrenstatik.* Springer, Berlin—Heidelberg—New York, 762pp.

Sherard, J. L., Woodward, R. J., Gizienski, S. F. and Clevenger, W. A., 1963. *Earth and Earth-Rock Dams.* Wiley, New York, N.Y., 725pp.

Striegler, W. and Werner, D., 1969. *Dammbau in Theorie und Praxis.* Springer, Vienna—New York, 462pp.

Walters, R. C. S., 1962. *Dam Geology.* Butterworth, London, 335pp.

Wegman, E., 1927. *The Design and Construction of Dams.* Wiley, New York, N.Y., 8th ed., 740pp. (with 111 plates).

Chapter 2

ROCKS AND UNCONSOLIDATED DEPOSITS: ORIGIN, COMPOSITION, AND PROPERTIES

INTRODUCTION

Geological and geophysical investigations of dam and reservoir sites and preparation of an inventory of materials available for construction of a dam and appurtenant features require careful petrographic and soil-mechanics studies of foundations and surficial deposits. The aim of these studies is classification and determination of properties that are of engineering concern, so that appropriate allowances can be made for the expected behavior of the various materials in the design and building of structures with acceptable margins of safety.

The emphasis in this chapter is on the geological classification of natural substances and the strengths of these substances as related to their origin, fabric, and mineralogy. No attempt is made to review or summarize the numerous engineering classifications of materials, especially of soils, that have grown out of the broad and expanding discipline of soil mechanics. The reader interested in the theory and application of soil mechanics is referred to the very extensive literature on the subject.

MINERAL COMPONENTS OF ROCKS AND UNCONSOLIDATED DEPOSITS

Rocks and unconsolidated deposits consist of aggregates of minerals of differing physical and chemical properties. Although thousands of minerals have been identified, only a few are found in significant amounts in rock bodies and in the secondary products derived from them. Although the science of mineralogy has reached an advanced stage of technical and theoretical sophistication, relatively few procedures are required for characterization of minerals as they contribute to the engineering properties of foundations of dams and reservoirs and as substances used in construction.

Properties of minerals apparent in hand specimen or under a hand lens are *color, luster, habit,* and *cleavage,* or the lack of it. Iron in minerals generally results in dark colors including gray, black, green, and brown. Luster is determined by the manner of reflection of light and is characterized as *nonmetallic, submetallic,* or *metallic.* Cleavage is a property which results in splitting upon impact along one or more planar surfaces and varies in quality from absent or highly imperfect, as in quartz, to nearly perfect, as in mica. *Habit* defines the shape commonly assumed by crystals or grains of a substance.

Other properties measured or estimated are specific gravity (density) and Moh's hardness (Table 2-1). Each mineral in the Moh's scale will scratch any one of

TABLE 2-1

Moh's scale of hardness

Reference mineral	Hardness	Reference mineral	Hardness
Talc	1	Feldspar	6
Gypsum	2	Quartz	7
Calcite	3	Topaz	8
Fluorite	4	Corundum	9
Apatite	5	Diamond	10
Comparison hardness:			
Fingernail	2+	Window glass	$5\frac{1}{2}$
Copper coin	3±	File steel	$6\frac{1}{2}$
Knife blade	5+		

lower number in the scale. With experience hardnesses of 5 and below can be estimated easily by scratching a substance with the point of a knife blade.

In additon to the use of the properties and simple tests mentioned above in the identification of minerals, the trained mineralogist has available to him a wide range of techniques for the study of minerals and their identification in single crystals and aggregates. Outstanding are X-ray diffraction methods and petrographic techniques employing the polarizing microscope.

Table 2-2 lists some properties of volumetrically important minerals. All of the minerals exhibit one or more cleavages of varying degrees of perfection except for limonite, olivine, garnet, chalcedony, quartz, hematite, and magnetite which do not possess visible cleavages in mineral grains.

The clay minerals play a special role in the engineering properties of natural substances. Because of their generally very small particle size (2 microns and less) their identification requires special techniques utilizing an X-ray diffractometer, a differential thermal analyzer, and chemical tests. Because of the platy nature and small size of particles, clay minerals in aggregates have a remarkable capacity to absorb and retain water and develop cohesion and plasticity. Some clay minerals, especially montmorillonite, also absorb water, and aggregates swell when moistened.

Clay minerals that mineralogically are classified as such are micaceous hydrous aluminum silicates. In some species magnesium and iron substitute in part for aluminum, and alkalies and alkali earths may be present in notable amounts.

TABLE 2-2

Some properties of volumetrically important minerals

Mineral	Composition	Hardness	Specific Gravity	Usual Habit
Talc	hydrous Mg silicate	1	2.7–2.8	micaceous
Gypsum	$CaSO_4 \cdot 2 H_2O$	2	2.32	tabular
Rock salt	NaCl	$2\frac{1}{2}$	2.16	equidimensional
Clay minerals	hydrous silicates	$2-2\frac{1}{2}$	2.6–2.7	micaceous; micro-crystalline
Chlorite	hydrous Mg,Fe,Al silicate	$2-2\frac{1}{2}$	2.6–2.9	micaceous
Mica				
Biotite	hydrous K,Mg,Fe Al silicate	$2\frac{1}{2}-3$	2.8–3.2	micaceous
Muscovite	hydrous, K,Al silicate	$2-2\frac{1}{2}$	2.8–3.1	micaceous
Limonite	hydrous iron oxide	$2-5\frac{1}{2}$	3.6–4.0	amorphous; micro-crystalline
Calcite	$CaCO_3$	3	2.71	equidimensional
Anhydrite	$CaSO_4$	$3-3\frac{1}{2}$	2.9–3.0	tabular; equidi-mensional
Dolomite	$CaMg(CO_3)_2$	$3\frac{1}{2}-4$	2.85	equidimensional
Serpentine	hydrous Mg silicate	$2-5$	2.6	microcrystalline; micaceous
Amphibole				
Tremolite	hydrous Ca,Mg silicate	5–6	3.0–3.3	prismatic
Hornblende	Hydrous Ca,Mg, Fe,Al silicate	5–6	3.2	prismatic
Kyanite	Al_2SiO_5	5–7	3.56–3.66	bladed
Pyroxene				
Enstatite	Mg,Fe silicate	$5\frac{1}{2}$	3.2–3.5	stubby, prismatic
Augite	Ca,Mg,Fe,Al silicate	5–6	3.2–3.4	stubby, prismatic
Hematite	Fe_2O_3	$5\frac{1}{2}-6\frac{1}{2}$	5.26	microcrystalline to tabular

Table 2-2 (continued)

Mineral	Composition	Hardness	Specific Gravity	Usual Habit
Magnetite	Fe_3O_4	6	5.18	equidimensional
Feldspar				
Orthoclase	K,Al silicate	6	2.57	equidimensional
Microcline	K,Al silicate	6	2.54–2.57	equidimensional
Soda plagio- clase	Na,Ca,Al silicate	6	2.62–2.69	stubby tabular
Calcium plagioclase	Ca,Na,Al silicate	6	2.71–2.74	tabular
Sillimanite	Al_2SiO_5	6–7	3.23	needles
Olivine	Mg,Fe silicate	$6\frac{1}{2}$–7	3.3–4.4	equidimensional to tabular
Garnet				
Almandine	Fe,Al silicate	7	4.25	equidimensional
Grossularite	Ca, Al silicate	$6\frac{1}{2}$	3.53	equidimensional
Andradite	Ca,Fe silicate	7	3.75	equidimensional
Chalcedony (chert)	SiO_2	$6\frac{1}{2}$–7	2.65	microcrystalline
Quartz	SiO_2	7	2.65	equidimensional

PROCESSES OF FORMATION OF ROCK BODIES

Rocks are natural aggregates of minerals which have sufficient cohesiveness to offer moderate to considerable resistance to mechanical breakdown. All gradations exist between rocks and unconsolidated materials such as alluvial deposits from streams. In a general sense, rocks are natural materials which constitute the earth's outer crust and formed as a consequence of processes active within the shallower to deeper portions of the crust at temperatures and pressures which are greater than those that prevail at the earth's surface.

A broad distinction is made among *igneous, sedimentary,* and *metamorphic* rocks. *Igneous rocks* have formed within or upon the surface of the earth's crust by the cooling and solidification of initially very hot silicate melts, called *magmas*. *Sedimentary rocks* result from the lithification after burial of accumulations of mechanically transported particles or substances precipitated from aqueous solutions. *Metamorphic rocks* are generated by the partial to complete recrystallization of previously existing rocks of all kinds at levels in the earth's crust where temperatures are sufficiently high to promote recrystallization but not melting.

The earth, the hydrosphere, and the atmosphere together constitute a dynamic system constantly undergoing change. Dislocations of the materials of the earth's interior and crust cause chemical imbalances, and processes are initiated which tend to restore physico-chemical equilibrium. Much of the earth is covered by water which prevents extensive chemical interaction between crustal rocks and the atmosphere, but where rocks are exposed at the surfaces of land areas, weathering and erosion promote extensive chemical and physical changes. An appreciation of the complexity of earth processes and the bewildering variety of rocks produced by them is gained by consideration of geochemical cycles.

In Fig.2-1 it is assumed that a geochemical cycle begins with weathering

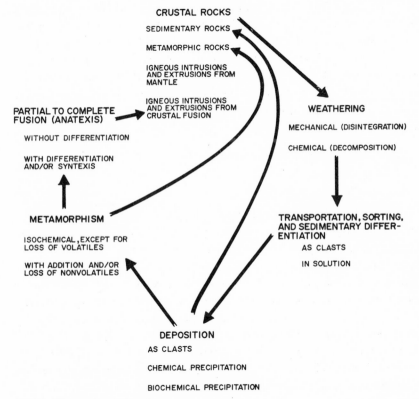

Fig.2-1. The geochemical cycle.

decomposition and disintegration of rocks exposed to the atmosphere at the earth's surface. The products of weathering are transported to sites of deposition and produce a variety of sedimentary rocks depending on the nature of the source rocks, the kind and intensity of weathering, the mode and distance of transportation, and the environment at the site of deposition. Sediments deposited as a veneer on the oceanic floor at great distances from continental masses are in effect removed from the geochemical cycle, but sedimentary accumulations on and near the margins of land masses remain in the cycle.

The cycle may be completed when buried sedimentary rocks are brought by crustal movement to the earth's surface, where, again, they are exposed to the atmosphere. Alternatively, displacement of the sediments downward into the hot interior of the earth produces metamorphic rocks which, under appropriate conditions of elevated temperature, may be partly to completely converted to magma. The cycle may be terminated with the elevation of the metamorphic rocks to shallower levels in the crust or by the intrusion and extrusion of magmas produced by fusion (anatexis).

Although they constitute only a small proportion of the rocks in continental sialic sectors, some igneous rocks, especially basaltic and peridotitic rocks, appear to have been derived from the upper portion of the mantle near the base of the crust. These rocks, especially where they have intruded older rocks, or as lava flows have spread out over the surface of the land or over the sea floor near continental margins, have the potential of becoming involved in the geochemical cycle in the same manner as the other, generally more silicic rocks with which they become associated.

IGNEOUS ROCKS

The common and volumetrically important igneous rocks consist dominantly of silicate minerals, especially feldspar and quartz. Abundant accessory minerals include mica, amphibole, pyroxene, and olivine. The grain size of igneous rocks is to a large extent a function of chemical composition and rate of cooling of the bodies of magma from which they congeal. Grain sizes reach a maximum in pegmatites where crystals ranging from a few centimeters to a meter or more in diameter are common. In contrast, quickly cooled igneous rocks sometimes have not crystallized at all and consist of glass, as in pitchstone and obsidian.

Igneous rocks as they rise from the depths of the earth as magmas invade or displace pre-existing rocks and upon solidification form bodies of a variety of shapes. Bodies that cut across structures of the pre-existing rocks are described as *discordant,* and those that parallel structures are classified as *concordant.* When magma reaches the earth's surface it flows over the surface as lava or is explosively ejected to develop fragmental *pyroclastic deposits.* In this book the pyroclastic deposits are classified with the sedimentary rocks.

A complete petrographic classification of the igneous rocks based on chemical analysis and microscopic examination is beyond the scope of the present treatment. Instead, a classification is presented in Table 2-3 which is based on features that can be observed in hand specimens and field exposures with or without the aid of a

TABLE 2-3

Key for megascopic classification of igneous rocks

I. Rock is glassy and
    A. Contains few or no phenocrysts:
         obsidian (vitreous luster)
         pitchstone (dull or waxy luster)
    B. Contains conspicuous or abundant phenocrysts:
         vitrophyre, either *pitchstone vitrophyre* or *obsidian vitrophyre*

II. At least 50% of rock is very fine-grained and outlines of crystals are below threshold of
    visibility.
    A. Rock is light-colored and entirely very fine-grained:
         felsite
    B. Rock is dark-colored and entirely very fine-grained:
         basalt or trap
    C. Rock is light-colored and contains abundant phenocrysts:
         1. Quartz and feldspar can be recognized in phenocrysts:
              rhyolite porphyry
         2. Feldspar, but no quartz can be recognized in phenocrysts:
              trachyte porphyry
    D. Rock is dark-colored and contains abundant phenocrysts of tabular feldspar:
         basalt porphyry

III. At least 50% of rock is coarse-grained enough to allow visual examination of individual
     grains.
     A. Rock is light-colored and approximately equigranular:
          1. Quartz and feldspar are abundant:
               granite
          2. Feldspar is abundant but no quartz is observed:
               syenite
     B. Rock is light-colored and porphyritic:
          1. Quartz and feldspar are abundant:
               granite porphyry
          2. Feldspar is abundant but quartz is absent:
               syenite porphyry

IV. Rock is medium-to-coarse-grained and contains abundant dark-colored accessory mine-
    rals. Feldspar is generally tabular and shows twin striations on cleavage surfaces:
         gabbro
         (anorthosite is a variety of gabbro consisting almost entirely of calcic plagio-
         clase)

V. Rock consists of very large (3 cm or more) crystals, commonly quartz and feldspar:
         pegmatite
         (sugary-grained quartz-feldspar aggregates commonly associated with peg-
         matite are called *aplite*)

VI. Rock consists largely of olivine and/or pyroxene:
         peridotite (abundant olivine)
         dunite (mostly olivine)
         pyroxenite (mostly pyroxene)

hand lens. In Table 2-3 the term *phenocrysts* refers to larger crystals set in a matrix or groundmass of glass or smaller crystals. Igneous rocks which contain phenocrysts are described as *porphyritic* if phenocrysts are few in number, or are classified as *porphyries* if phenocrysts are abundant or especially conspicuous.

SEDIMENTARY ROCKS

Sedimentary rocks form by the accumulation of fragments (clasts) of rocks or minerals transported to a site of deposition by water, wind, or ice and by the accumulation of substances precipitated from chemical solutions. Initially accumulations are unconsolidated and consist of loosely aggregated particles, but, with burial and time, changes occur, especially compaction and development of interstitial cement, that result in *lithification*.

Sedimentary rocks can be classified into two broad categories (1) dominantly clastic rocks held together with a compacted matrix or a chemical cement such a silica, carbonates, or iron oxides, and (2) dominantly crystalline rocks such as rock salt, gypsum, anhydrite, and certain crystalline limestones and dolostones, which were constituted into crystalline aggregates at the time of, or slightly later than, their accumulation in strata, or which recrystallized at a considerably later time owing to deep burial and/or activity of groundwater.

Table 2-4 is a key for megascopic identification of common sedimentary

TABLE 2-4

Key for megascopic identification of sedimentary rocks

I.   Rock is dominantly an accumulation of clastic silicate particles or rocks bonded by a matrix and/or a cement:

      A. Average size of particles is clay- or silt-size. Clay-size particles are below the threshold of visibility in hand specimen and even under the microscope. Silt-size particles are below or just at the threshold of visibility with a hand lens, but are so small that identification is not generally possible.

         claystone
         mudstone
         shale (bedded and fissile)
         siltstone

      B. Average size of clastic particles is sand-size (about 0.06–2 mm):

         sandstone

        1. Particles are angular and consist of a variety of minerals, especially feldspar and quartz. Matrix contains clay minerals.

         graywacke sandstone

        2. Particles are angular and consist mainly of feldspar and quartz. Clay matrix is absent.

         arkosic sandstone

TABLE 2-4 (continued)

---

        3. Particles are rounded and consist mainly of quartz. Cement is generally silica or carbonate.
           quartzose sandstone
    C. Average size of particles is greater than 2.0 mm or conspicuously larger rock or mineral fragments are present in a finer-grained aggregate:
        1. Particles are angular:
           conglomerate breccia
        2. Particles are more or less rounded:
           conglomerate

II.   Rock is dominantly an accumulation of clastic particles of carbonate minerals:
        1. Very fine-grained:
           micritic limestone (consists of tiny calcite particles and generally requires microscopic examination before appropriate identification)
        2. Particles are of sand size (0.06–2 mm):
           calcarenite (calcite)
           doloarenite (dolomite)
        3. Average size of particles is greater than 2.0 mm or conspicuously larger fragments are present in a finer-grained aggregate:
           limestone breccia
           dolomite breccia
           limestone conglomerate
           dolomite conglomerate

III.   Rock consists dominantly of anhedral to euhedral crystals and fabric is nonclastic.
           crystalline limestone (calcite)
           crystalline dolostone (dolomite)
           gypsum
           anhydrite
           rock salt
IV.   Rock is pyroclastic, having formed by accumulation of particles of explosive volcanic origin, crystals and/or glass.
        Particles greater than 32 mm in diameter:
           volcanic breccia; volcanic agglomerate
        Particles 32–4 mm in diameter:
           lapilli tuff
         Particles $4-\frac{1}{4}$ mm in diameter:
           coarse tuff
         Particles less than $\frac{1}{4}$ mm in diameter:
           tuff; if mainly glassy: *vitric tuff*

---

rocks. Rocks of mixed fabric or composition can be further classified by using appropriate modifiers, for example, *calcareous shale, sandy shale, clayey sandstone,* etc.

METAMORPHIC ROCKS

Metamorphic rocks have formed by partial to complete recrystallization of pre-existing rocks at high temperatures and/or pressures. Depending on the intensity of metamorphism, distinctions are made among low-grade, medium-grade, and high-grade mineral assemblages. Under many conditions metamorphic rocks are notably deformed during recrystallization and assume very complex, highly contorted fabrics. Layering, either inherited from the previous condition of the rock or generated during metamorphism is called foliation or schistosity, and is conspicuously displayed in the great preponderance of metamorphic rocks.

A classification based on features observed in hand specimens and, especially, in field exposures, is given in Table 2-5.

TABLE 2-5

Outline classification of metamorphic rocks

I.   Mineral components are randomly oriented in hand specimen, whether platy, elongate, or equidimensional. Foliation not apparent or very poorly developed.
        quartzite (mainly quartz)
        marble (mainly calcite)
        dolomitic marble (mainly dolomite)
        hornfels (fine-to-coarse-grained, mainly silicate minerals)

II.   Rock consists mainly of platy, micaceous, prismatic, or acicular minerals in subparallel to parallel arrangement.
    1. Very fine-grained. Grains below limit of visibility in hand specimen. Rock generally splits easily into thin plates.
       slate
    2. Fine-grained. Some grains can be seen with hand lens. Splitting tendency often pronounced.
       phyllite
    3. Particles are large enough to be observed in hand specimen.
       schist, especially *biotite schist, muscovite schist,* and *hornblende schist*

III. Rock contains abundant quartz and/or feldspar and layering or linear arrangement of component minerals is not especially pronounced as in schist. In some rocks and on a small scale layers rich in *quartz* and/or *feldspar* alternate with layers rich in platy, micaceous, prismatic, or acicular minerals.
       gneiss, especially *biotite gneiss, muscovite gneiss,* and *hornblende gneiss*

IV. Rock appears to be a complex intermixture of metamorphic rocks and granular igneous rocks.
       migmatite

DISINTEGRATION AND DECOMPOSITION OF ROCKS

Rocks are subject to a variety of physical and chemical processes which alter their properties in varying degrees. Generally alterations result in a notable reduction in grain size and, concomitantly, in a reduction in strength. The most familiar type of alteration is weathering as a consequence of exposure of rocks to the atmosphere, but many rocks are pervasively changed by penetration of hot aqueous solutions (hydrothermal solutions) of deep-seated origin.

The rate of weathering of a rock body depends on its composition, the climate, and the extent to which surface solutions can gain access along fractures or other avenues of groundwater circulation. Maximum rates of weathering are noted in warm to hot regions of high rainfall. Minimum rates of weathering are observed in arid or cold regions. Weathering, together with organisms, produces soils, and conditions rocks for easy erosion by wind, running water, or moving glacial ice.

The products of chemical weathering are diverse. Iron released by weathering combines with oxygen and/or water to be precipitated as hematite (Fe_2O_3) or limonite ($Fe_2O_3 \cdot nH_2O$). Calcium and magnesium combine with carbon dioxide to form calcite ($CaCO_3$) and dolomite ($CaMg(CO_3)_2$), or, in poorly drained environments, become components of swelling clays. Aluminum usually becomes incorporated in clay minerals. The alkalies, sodium and potassium, are leached or enter into clay minerals.

Table 2-6 indicates the varying degrees of resistance of minerals to chemical

TABLE 2-6

Resistance of silicate minerals to chemical weathering

1. Low resistance	
olivine	calcic plagioclase
pyroxene	biotite mica
hornblende	iron-rich garnet
2. Intermediate resistance	
intermediate plagioclase	muscovite mica
sodic plagioclase	sillimanite
orthoclase feldspar	kyanite
microcline feldspar	iron-poor garnet
3. High resistance	
quartz	hematite
magnetite	clay minerals

weathering and suggests the reasons for greater resistance to weathering of some rocks as compared with others. In general rocks containing abundant magnesium, calcium, and iron breakdown chemically by weathering alteration more easily than rocks deficient in these elements.

Hydrothermal solutions produce a number of kinds of alterations by introduction of certain chemical elements and leaching removal of others. As in weathering, hydrothermal alteration commonly converts rocks into finer-grained, weaker aggregates. Some kinds of hydrothermal alteration are as follows:

(1) *Argillization,* with development of clay minerals.

(2) *Sericitization,* with introduction of fine-grained muscovite mica (sericite).

(3) *Silicification,* by introduction of small to large crystals of quartz. Commonly silicification improves the strength of an initially weak rock.

(4) *Propylitization,* characterized by formation of chlorite, clay minerals, and carbonate minerals at the expense of pre-existing minerals. Usually observed in fine-grained igneous intrusions, lava flows, or pyroclastic deposits.

(5) *Chloritization,* with the development of chlorite from pre-existing silicate minerals containing magnesium, iron, and aluminum.

(6) *Dolomitization,* by the introduction of dolomite ($CaMg(CO_3)_2$), usually into limestone.

(7) *Carbonatization,* with the introduction of calcite ($CaCO_3$).

UNCONSOLIDATED DEPOSITS

Unconsolidated deposits form a mantle covering most of the earth's surface. Weathering by mechanical and chemical processes commonly produces aggregates with varying degrees of cohesion, ranging from practically none to a degree of cohesion approaching that of the original material. Development of soils capable of supporting the growth of organisms further reduces cohesion, except where chemical compounds such as iron oxides and carbonates, are precipitated as cement by moving, near-surface solutions.

Rocks conditioned by weathering tend to move downslope under the influence of gravity and by expansion and contraction to accumulate locally as *colluvial deposits.* Familiar examples are talus piles, landslides, and soil-creep accumulations.

Many unconsolidated deposits have been transported to their present locations by moving water, wind, or glacial ice from more or less distant sources. Included are wind-carried *loess* and *sand-dunes,* alluvial *stream deposits, glacial till,* and *fluvioglacial deposits* transported by melt waters from melting glacial ice. The characteristics of the deposits are a function of the source materials and the mechanism and intensity of the transporting process. Very few transported deposits are uniform throughout and record cyclic and noncyclic changes in the mechanisms and environments of deposition.

Unconsolidated deposits through burial tend to become denser and more cohesive. In some deposits precipitation of interstitial cement converts loose aggregates into firm, rock-like masses.

The particular classification of unconsolidated materials that may be employed depends on the purpose of the classification. In a broad sense a distinction is made between geological classifications (petrographical classifications) and engineering classifications. In geological classifications particular attention is paid to the shapes, sizes, and mineralogy of the various components. In engineering classifications emphasis is placed on the properties of unconsolidated materials as they relate to their behavior in foundations or excavated slopes, and as materials for construction. Engineering classification of unconsolidated aggregates is beyond the scope of the present treatment. The interested reader will find a vast amount of information concerning engineering classifications in many current publications on soil mechanics. A particularly useful reference is the *Earth Manual* (U.S. Bureau of Reclamation, 1968).

In a geological classification a scale of grain size for unconsolidated deposits is given in Table 2-7. Grain sizes are measured by various techniques including use of

TABLE 2-7
Grain-size classification of unconsolidated deposits

Designation	Particle diameter (mm)
Cobbles, bowlders, blocks	> 60
Coarse gravel	60–20
Medium gravel	20–6
Fine gravel	6–2
Coarse sand	2–0.6
Medium sand	0.6–0.2
Fine sand	0.2–0.06
Coarse silt	0.06–0.02
Medium silt	0.02–0.006
Fine silt	0.006–0.002
Clay	< 0.002

nested calibrated screens, air and water elutriators, and hydrometers. Mineral composition is usually determined by a combination of methods, especially the petrographic technique utilizing a polarizing microscope and, for particles of very small size, the X-ray diffractometer technique. Depending on the distribution of various particle sizes in deposits, unconsolidated deposits are characterized as very poorly sorted, poorly-sorted, well-sorted, or very-well sorted. Examples of poorly-sorted accumulations are clayey sandstones and glacial till. Well-sorted deposits are exemplified by dune sand, beach sand, and stream sand from which the very fine-grained particles have been removed.

The mineral components of natural unconsolidated deposits vary widely in particle size and mineralogy. In engineering practice unconsolidated deposits are generally included under the broad term *soil*. *Soil mechanics* is an extensive discipline of engineering beyond the scope of this book. Ordinarily the geologist does not become involved in the more or less complicated procedures of soil testing for engineering purposes, whether for embankment construction or as dam foundations. However, geologists and geophysicists frequently play an important role in locating and estimating volumes of materials that might prove to be suitable in embankment construction or for concrete aggregate.

The geologist is at hand during core drilling, geophysical testing, and permeability testing of foundations in unconsolidated deposits and assists in the interpretation of vertical and lateral inhomogeneities in the deposits, particularly as they may contribute to seepage beneath a dam. An additional contribution to an understanding of the behavior and properties may be a laboratory investigation of the detailed mineralogy of a particular soil and an assessment of the probable long-term behavior of the soil under pressure from a dam and reservoir and when penetrated by seepage water.

ROCK FABRIC AND COMPOSITION RELATED TO STRENGTH

Although loose aggregates of sandy or gravelly materials possess no strength unless confined, many clayey accumulations have measurable shear strength. So-called *firm, stiff,* and *hard clays* are sufficiently cohesive that forces of between 5 and 20 psi are required to cause shear failure. For clays of similar mineralogical composition and particle size an increase in pore water is accompanied by a decrease in cohesiveness.

Penetration of loose aggregates by groundwater which deposits cement or causes partial to complete recrystallization of existing components increases strength and cohesiveness, and, by degrees, initially unconsolidated deposits may gradually assume the characteristics of materials classified as rocks.

The strength of unfractured rocks, as measured by crushing strengths determined in the field or laboratory, is a function of *mineralogy* and *rock fabric.* The term *fabric* refers to the three-dimensional aspects of rocks as expressed by grain size, grain shape, grain distribution, and the manner of articulation of mineral particles. A distinction is made between fabrics observed in hand specimens of field exposures, *megafabrics,* and fabrics that are apparent only on microscopic examination, *microfabrics.*

Fabrics also are classified as *isotropic* and *anisotropic.* In isotropic fabrics rock properties are the same in all directions. Anisotropic fabrics display linear or planar arrangements of elements of fabric and are exemplified by stratification in sedimentary rocks, foliation and schistosity in metamorphic rocks, and linear parallelism of prismatic minerals such as amphibole, as seen in some igneous and metamorphic rocks.

From an engineering point of view, where strength is a primary consideration, petrographic classification, although useful in supplying a name for a rock, is not as pertinent as the response of the rock to loads, or its properties when used as a construction material.

A basic difference in strength of rocks as related to fabric is observed in rocks which have resulted from the layer-upon-layer accumulation of particles (clasts) by

· sedimentary processes and in rocks which have partly or completely recrystallized from pre-existing rocks or have crystallized from cooling magmas. In the latter interlocking articulation of mineral grains generally provides much greater strength than observed in many sedimentary accumulations. Accordingly, in following discussions a distinction will be made between *clastic rocks* and *nonclastic* (crystalline) *rocks* and the processes of formation and fabrics that characterize each.

STRENGTH OF CLASTIC ROCKS

Clastic rocks are lithified accumulations of mineral and/or rock particles which have formed by sedimentary processes. Included among the clastic rocks are materials that were transported to the site of deposition by wind, water, or ice from a more or less distant source. Sites of deposition are on the land's surface (subaerial deposits), within channels or on flood plains of streams and rivers (alluvial deposits), or in standing bodies of water (lacustrine and marine deposits). Mechanically transported materials build up accumulations when the energy of the transporting medium is reduced below a critical level and particles dragged by moving currents or carried in suspension come to rest on a floor of deposition.

Transport of chemically dissolved substances followed by precipitation by inorganic processes or through the intervention of organisms results in layer-upon-layer accumulation of crystals or shells of organisms. Commonly precipitated substances are prone to repeated resolution and reprecipitation after deposition, and, with time, initially clastic fabrics are converted to crystalline fabrics as in deposits of rock salt, gypsum, and some limestones.

Burial of unconsolidated clastic accumulations eventually results in lithification by a complex process involving compaction, loss of interstitial pore water, and chemical reactions. The process whereby loose, unconsolidated aggregates are converted to rocks is called *diagenesis*.

Although they are of igneous origin, clastic deposits resulting from explosive eruption of volcanoes are sedimentary in that they have been transported by wind or water to the site of deposition, or have formed by downslope movement under the force of gravity of volcanic ejecta in over-steepened slopes adjacent to centers of eruption. These rocks, because of their particular origin, are classified as *pyroclastic rocks.*

Like other kinds of sedimentary rocks pyroclastic rocks may initially have consisted of loose aggregates of rock and mineral fragments, but unlike typical sedimentary rocks, the rock and mineral fragments may have been very hot at the time of accumulation. Quickly congealed lava usually is very fine-grained or glassy, and burial of hot crystal fragments and glassy fragments (shards) promotes welding and development of considerable cohesive strength. Welded aggregates associated with volcanism are described by the general terms *welded tuffs* or *ignimbrites.* Of course, accumulations of fragmental materials from explosive volcanoes at some distance from the source and deposited at low to moderate temperatures upon burial and with time are lithified by processes similar to those that affect other sedimentary accumulations.

In clastic rocks particles that are in contact with each other sufficiently to support a loose aggregate formed by the original process of sedimentation are *framework clasts.* The interstices among framework clasts are open spaces that may be variously occupied by fluids or smaller particles that constitute a *matrix.* An example is a sandstone with a very fine-grained clay-rich matrix filling voids among larger sand grains that are in physical contact with each other. Materials chemically precipitated in the voids simultaneously with initial deposition or a later time constitute a *cement,* which may be introduced into an aggregate from an outside source by percolating solutions or may develop by local solution and reprecipitation of matrix materials or framework clasts.

In shales the distinction between matrix and framework clasts is not so obvious as in coarser grained clastic rocks. Nevertheless, lithification of unconsolidated clay deposits to generate shale is accomplished by the same processes that involve other clastic rocks, that is, by introduction of materials from outside sources or by internal chemical readjustments. Many shales prior to lithification contain very fine-grained, even colloidal, components that are subject to easy chemical attack promoting nearly spontaneous chemical reactions.

A notable consequence of the formation of cement in clastic rocks is a dramatic increase in strength. Sedimentary rocks with pervasive, void-filling cement of silica, carbonates, and iron-oxides generally have high to maximum strengths and are much stronger (much more cohesive) than clastic rocks in which voids among framework clasts are filled with an unrecrystallized matrix.

Some sedimentary accumulations are particularly subject to self-cementation whereby solution at the contacts of framework clasts and reprecipitation in adjacent voids produces a cement of the same chemical and mineralogical composition as the framework clasts. Outstanding examples are highly quartzose sandstones (orthoquartzites) and many limestones.

Mineralogical composition plays an important part in determining strength even though a binding, interstitial cement is present. Minerals such as quartz and feldspar, or particles of rock fragments, common constituents of clastic rocks, have much higher inherent strengths than minerals such as the clay minerals, micas, and gypsum which are soft, tabular, or platy, and possess excellent to perfect cleavages. In rocks in which the latter minerals are abundant and in parallel to subparallel orientation strength is low because of a tendency to easy splitting parallel to the mineral cleavages. Examples of rocks with very low strengths owing to parallelism

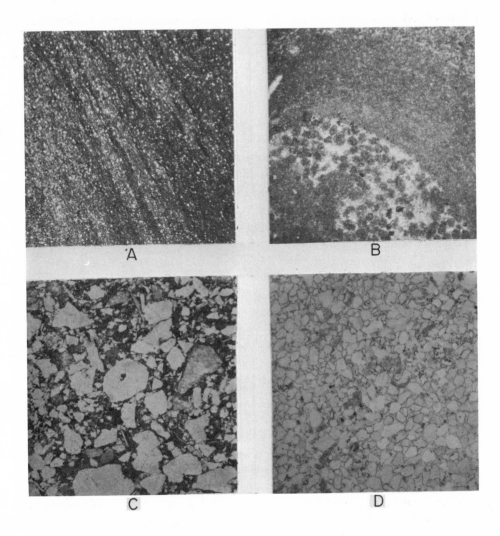

Fig.2-2. Photomicrographs of petrographic thin sections of clastic sedimentary rocks.

A. Sandy shale. A low-strength rock consisting mainly of clay minerals and tiny quartz grains.

B. Fine-grained limestone (micrite) containing microfossils. A high-strength, compact rock.

C. A sandstone with angular fragments of quartz and feldspar embedded in a clay matrix. Low to moderate strength.

D. A sandstone with siliceous cement (orthoquartzite). A high-strength rock.

or subparallelism of clay minerals and/or micas are poorly cemented bedded shales and siltstones.

Additional factors determining the strengths of clastic rocks are grain size and geologic age. Clay minerals and micas not only are soft minerals with highly developed cleavages and if they are abundant, tend to produce weak rocks without regard to orientation or grain size. However, other minerals contribute to rock strength in a manner depending on grain size and, in general, fine-grained rocks have higher strengths than coarse-grained rocks of the same mineral composition.

It is well known that geologically old clastic rocks commonly have higher strengths than their younger counterparts, presumably because of compaction, more or less recrystallization, and formation of cement through prolonged periods of burial.

The wide range in crushing strengths of some common clastic rocks as a function of differing fabrics and mineralogies is indicated in Table 2-8. Examination

TABLE 2-8

Unconfined crushing strengths of some common clastic rocks

Rock	Crushing strength (psi) *
Calcareous mudstone	8,000–28,000
Dolostone (dolomite)	9,000–51,000
Limestone	700–29,000
Sandstone	1,500–34,000
Shale	1,000–33,000
Siltstone	4,000–45,000

*To convert psi to bars divide by 14.5.

of this table clearly indicates that strength does not correlate exlusively with a particular category of rocks. However, in a general characterization clastic rocks in order of increasing strength are as follows: shale (weakest), siltstone, sandstone with minor cement, limestone, dolomite, and quartz sandstone with siliceous cement (strongest).

Examples of clastic rocks with varying degrees of strength are shown in Fig.2-2.

STRENGTH OF CRYSTALLINE (NONCLASTIC) ROCKS

The crystalline rocks include igneous rocks (except the glassy igneous rocks), metamorphic rocks, and sedimentary rocks that have recrystallized from a previous clastic condition. Crystalline sedimentary rocks are exemplified by most deposits of rock salt, gypsum, and anhydrite and many accumulations of limestone and dolomite. All share fabrics characterized by interlocking crystals which are in close contact with each other along straight to sinuous boundaries. All of the crystalline rocks generally have high strengths, except rock salt and gypsum, which consist of inherently weak, soft minerals with excellent cleavages.

The strengths of metamorphic rocks vary widely, depending on the extent of recrystallization of source rocks and the amount and distribution of the micaceous minerals, biotite, muscovite, and chlorite. In most metamorphic rocks foliation or

TABLE 2-9

Ranges of unconfined crushing strengths of some common unaltered igneous and metamorphic rocks

Rock	Crushing strength (psi) *
Basalt	26,000—40,000
Diabase	22,000—46,000
Felsite and felsite porphyry	18,000—38,000
Gneiss	22,000—36,000
Granite	6,000—42,000
Marble	7,000—34,000
Phyllite (micaceous)	1,000— 2,500
Quartzite	30,000—53,000
Schist, all varieties	1,100—20,000
Schist, mica-rich	1,100— 7,200
Slate	14,000—47,000

*To convert to bars divide by 14.5.

A   B

Fig.2-3. Photomicrographs of petrographic thin sections of igneous rocks.
A. Felsite porphyry. A high-strength rock consisting of phenocrysts of feldspar embedded in a microcrystalline groundmass.
B. Syenite consisting mostly of feldspar. A high-strength rock with interlocking mineral grains. Photograph in polarized light.

schistosity is visible in hand specimens and field exposures as a consequence of
segregation of minerals of different sizes and/or compositions into adjacent layers.
Strengths measured by compression at right angles to layering generally are greater
than those measured by pressures acting parallel to the layering.

Among the weakest of the metamorphic rocks are mica or chlorite-rich slate,
phyllite, and schist. Among the strongest rocks are those in which mica is absent or

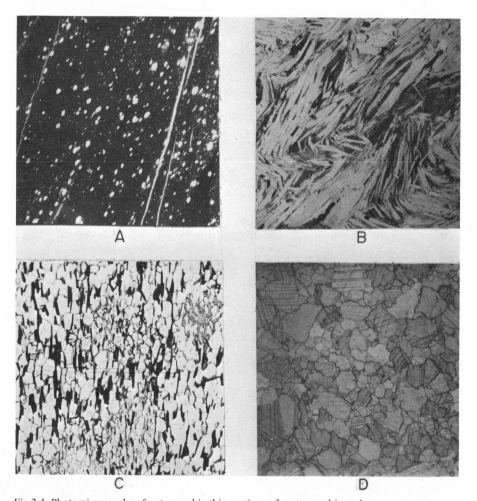

Fig.2-4. Photomicrographs of petrographic thin-sections of metamorphic rocks.
A. Microcrystalline slate containing scattered quartz particles. A moderate- to high-strength
rock. Note cracks that have formed along a slaty cleavage.
B. Muscovite schist. A low-strength rock containing contorted layers of mica.
C. Biotite gneiss. A high-strength rock containing abundant quartz and feldspar and minor
biotite mica.
D. Marble. A very high-strength rock consisting of interlocking calcite grains.

present only in small amounts. Many gneisses consist of strong layers of quartz and/or feldspar alternating with layers of mica of low splitting strength. In large bodies the bulk strength of these gneisses is determined by the spacing and thickness of the mica-rich layers.

Some values for unconfined crushing strengths for several igneous and metamorphic rocks are given in Table 2-9. As for clastic rocks a wide range is apparent. Variations are attributable to variations in fabric and mineralogy. Again, as for clastic rocks, an abundance of mica, whether in tiny flakes or larger crystals, correlates with low strengths.

No distinction is made in Table 2-9 between fresh and altered rocks. Alteration with the formation of fine-grained clay minerals greatly reduces the strength of rocks.

Examples of igneous and metamorphic rocks of varying strengths are shown in Fig.2-3 and 2-4.

REFERENCE

U.S. Bureau of Reclamation, 1968. *Earth Manual.* 1st ed., revised.

EXCAVATION AND FILLING OF VALLEYS

INTRODUCTION

A *valley* is an elongate, commonly sinuous stretch of lowland lying between hills, mountains, or bluffs and usually contains a stream or river which flows through it. A *canyon* is a deep narrow valley between high vertical cliffs or steep slopes. The terms *gorge, defile, chasm,* and *glen* also are used for narrow, deep valleys. A *ravine* is a small valley cut by running water. Smaller topographic features characterized by elongate channels of relatively shallow depth are *gullies, washes,* and *arroyos*, all of which generally contain only intermittent streams.

Valleys, large and small, have formed or have been modified by downward and lateral erosion of running water and/or ice, and commonly contain unconsolidated deposits transported by water, ice, or wind, and products of downslope movements under the influence of gravity of soils and products of mass wasting. The individual characteristics of a valley are a function of the topography, climate, rock type, and geologic structure as they exist now or have existed in the relatively recent geologic past.

Many valleys have long and complicated histories, and features observed today may have their origins in the recent or distant geologic past. Some of the processes that are related to stream cutting and valley development are local fold and/or fault development, elevation or depression of broad segments of the earth's crust, damming of drainages by massive earth slides or volcanic deposits, changes in climate, and changes in sea level.

This chapter is concerned with valleys of a variety of aspects in essentially homogeneous materials, and attention is given to processes and products of erosion and deposition by streams and glaciers whose erosional activities are not greatly influenced by structural anisotropism in bedrocks. Discussion of the influence of geologic structure in the development of erosional and depositional features in valleys is presented in a subsequent chapter.

Except for reservoirs constructed with embankments on three or four sides, as in water-storage reservoirs within or near municipalities, artificial reservoirs usually are created by construction of a dam or dams in a large or small valley, commonly in a constriction. From an engineering and geological standpoint a knowledge of the origin and nature of valleys and the origin and characteristics of the unconsolidated materials on the floor and slopes of valleys is of prime concern in locating and planning the safe construction of dams and reservoirs. Correct interpretation of the various physical aspects of a valley usually reveals much concerning the characteristics of bedrock beneath a dam site and beneath the floor and sides of the reservoir basin above the dam site.

EROSION, TRANSPORTATION, AND DEPOSITION BY RUNNING WATER

Running water erodes the materials in the bottom and sides of the channel by corrosion, corrasion, and cavitation. *Corrosion* is a chemical process whereby materials are taken into solution so as to become part of the *dissolved load* of a stream. Limestone is particularly susceptible to this process. *Corrasion* is a mechanical process that causes materials to wear away and includes *abrasion* by solid particles carried by the stream, and *evorsion,* which wears down compact materials by the impact of clear water carrying no suspended load. Removal of loose particles by the force of moving water sometimes is called *hydraulicking.*

Cavitation requires high velocities in running water and results first from formation of vapor bubbles because of pressure decrease associated with velocity increase in accordance with the Bernoulli theorem and, then, explosive collapse of the bubbles where the velocity diminishes. The impact of energy pulses resulting from collapsing bubbles can be very destructive of nearby solid substances. Cavitation is a process that is probably active in streams only in waterfalls and rapids.

In swiftly moving streams in steep canyons erosion by abrasion and, perhaps, cavitation is capable of producing deeply excavated sinuous channels and large and small potholes in a very resistant bedrock such as granite or quartzite.

The *load* of a stream is the sum of the dissolved materials, the *dissolved load,* and the *solid load.* Forward movement, lateral and vertical movements, turbulence, vortices, and eddys work together to support the *suspended load,* which, depending on stream velocity and the depth and shape of the channelway, range in size from clay-size particles to much larger particles. The *bed load* or *traction load* moves along the bottom of the channelway by rolling, slipping, or saltation. The last process is characterized by intermittent "jumping" of particles when the lifting force of the moving water exceeds the weight of a small particle, a pebble, or a bowlder.

Deposition of the solid load is a consequence of a decrease in a stream of gradient, volume, or velocity, or addition to the stream of debris from an outside source such as a rapidly moving tributary stream. Deposition, then, is the result of a decrease in transporting ability for whatever reason.

Among the most common depositional features of a stream, aside from local deposits in steep canyons, are *alluvial flood plains, deltaic deposits* in standing bodies of water such as lakes along the course of the stream, and *alluvial fans,* which accumulate where swift, heavily loaded streams flow out on flat areas.

It is evident that a stream valley exists only because downward and lateral erosion throughout the history of the valley have exceeded deposition of the load transported by the contained stream from upper reaches of the valley. However, at any given time and at a given place and largely as a function of the volume and velocity of water flowing through a valley, the average rate of erosion exceeds the average rate of deposition, or the reverse. A stream which, throughout its length and over a long period of time, has the capacity to transport its total load is said to be *graded.*

When considering the construction of a dam and reservoir in a valley the concern generally is with only a relatively short segment of the total length of a stream, and particular attention is given to whether in the floor of the valley erosional features on the average dominate or are subsidiary to depositional features. Also of vital importance are the characteristics and volume of the load transported by the stream into the reservoir and that may eventually fill it and render it useless as a storage facility.

In a typical valley maximum erosion and transportation of load occur during periods of flooding characterized by large volumes of swiftly flowing water, and load deposition or maintenance of an essentially static condition with little or no erosion and slight deposition are the rule during intervals of low water. However, even at times of most active erosion there is local deposition of the load.

Except for short lengths the courses of most streams are sinuous or meandering, even in valleys with straight, parallel banks or walls. The line connecting the deepest points in a channel, the *thalweg,* usually is winding although the stream flows between straight banks at the surface. That is, although the banks may be straight, the flow between them is not.

Flow of water in a channel is a combination of forward movement, surface water moving toward the center of the stream and downward, and movement of water from the bottom outward toward the walls and upward. The combination of these movements results in a spiral or helical movement of the whole mass of water. Irregularities in the channel, or obstructions, introduce a variety of complications in the total movement picture.

Fig.3-1 indicates diagrammatically some of the aspects of stream erosion and deposition in a winding channel. Fig.3-1A indicates the patterns of flow in the stream as observed from above. Fig.3-1B shows areas of deposition and areas of erosion by scouring of the channel during high flow. During high flow a *crossing bar* tends to develop diagonally across the channel, and erosion undercuts banks and deepens the channel at the indicated locations. Fig.3-1C indicates conditions during low flow when the crossing bar is eliminated or reduced and deposition causes growth of *point bars.* The location of the thread of maximum velocity shifts with the volume and velocity of the water in the channel so that the exact sites of erosion and deposition undergo constant change as flows increase and decrease. The

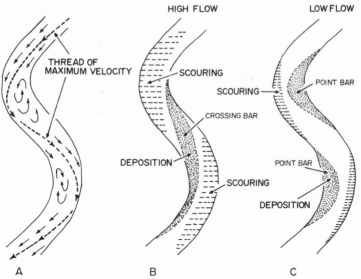

Fig.3-1. Erosion and deposition by a sinuous stream.

picture is further complicated at times of high floods when the stream overflows
the banks of the channel and deposits part of its load outside the channel.

Along intervals of streams where gradients are low, as on floodplains, a
natural consequence of the processes indicated in Fig.3-1 is the development of
meandering streams and lateral and downstream shifting of the stream curves in the
channels. In many floodplains the materials that are being eroded and transported
are unconsolidated sediments deposited on and above the bedrock excavated during
earlier periods of stream activity.

In regions of considerable topographic relief the processes indicated in Fig.3-1
tend to increase channel sinuosity simultaneously with downward cutting and local,
shifting deposition of the stream load. A net result is usually a highly irregular
longitudinal profile of the channel bottom along the thalweg.

CHANNEL PATTERNS AND CROSS-SECTIONS IN STREAM-CUT VALLEYS

Development of a valley of a particular plan configuration and cross-section
by a stream is a process which is sensitive to a variety of controls. The process
essentially is one of more or less balanced downward cutting and lateral widening
either by sideward erosion by the stream and/or sudden collapse or slow mass
movement of the materials in the valley walls toward the floor of the valley or
stream channel. Erosion, transportation of the bed load, and removal of products of
downslope movement of materials from the valley walls are at a maximum during

periods of high water and flooding, and many or most of the erosional features of valleys owe their origins to intermittent floods.

Many major stream drainages, particularly those of subcontinental dimensions, reveal features that indicate periods of flooding and erosion in the past that exceeded anything in the memory of man by several orders of magnitude. Over much of the face of the earth valleys and deep canyons hundreds to thousands of feet deep can not reasonably be explained as the result of erosion by the streams and rivers that presently flow through them. In the lower reaches of most large drainage basins tremendous volumes of alluvial material fill and bury deep or broad, shallow valleys cut into bedrock, and, where the rivers have entered the sea, large deltas have formed.

Spectacular downward cutting by running water in relatively recent geological times in many areas appears to be a complex function of uplift of large segments of the earth's crust by mountain building, crustal warping, post-Glacial rebound, and a changing climate that resulted in the repeated advance and recession of ice sheets, ice-caps, and valley glaciers during the ice ages.

Post-Glacial rebound is a phenomenon associated with the melting of the thick ice sheets that buried vast land areas, predominantly above 40° of latitude, during the ice ages. Removal by melting of the ice load, which elastically depressed the basement rocks was accompanied by rebound which, in some areas, elevated the depressed areas several hundreds of feet. For example, in the northern coastal areas of Canada and, as a consequence of rebound, marine shells and ancient beaches are found at heights in excess of 900 ft above present sea level.

Growth of the huge ice sheets and widespread valley glaciers required freezing of very large volumes of water. It has been estimated that less than 20,000 years ago, during the last great epoch of glaciation, the accumulation of ice in glaciers lowered worldwide sea level between 300 and 400 ft. Lowering of sea level during buildup of ice caps and valley glaciers in mountainous regions promoted erosion because of a lowered base level of erosion, but the intensity of stream erosion and transport of stream load must have been at a peak when glaciers retreated by melting, and great torrents of water were returned to the sea.

Whether the characteristics of a particular valley that we observe now are presently developing or owe their origins to events in the past, the valley shape in cross-section and in plan results from an interplay of climate, grade, and rock types within the valley and in its upper reaches.

Climate determines the volume of flow and the amount of vegetative cover that, when abundant as in wet hot climates, tend to provide a dense protective cover that prevents or inhibits erosion of valley walls. In distinct contrast, as in arid and semiarid regions, sparse vegetative cover leaves materials on the valley walls exposed to downslope movement and rapid erosion by intermittent streams and sheet wash coursing down slopes. Climate, coupled with rock type, determines the nature and intensity of weathering and soil development. There are important differences in the erodability of weathered rocks where solutions of surface origin by chemical weathering have reduced original rocks to loose, incoherent aggregates rich in clay minerals and rocks in arid or cold climates where the weathering breakdown is dominantly mechanical and produces only accumulations of angular rock rubble.

The grade of the stream through a particular interval of a valley determines the velocity and the cutting and transporting power of the stream occupying the valley. With a steep grade downcutting exceeds side cutting, and a V-shaped valley is a consequence. When grades are flatter, erosion is mainly lateral and tends to widen the floor of the valley, sometimes with the development of flat-floored, steep-walled, U-shaped cross-sections.

Either downcutting or lateral erosion are more rapid in unconsolidated materials such as gravel and clay and soft sedimentary rocks such as siltstone, shale, and mudstone than in hard, corrasion-resistant rocks including granite, quartzite, limestone, and most metamorphic rocks. Because of these differences deep steep-walled canyons with narrow stream channels develop only in rocks which strongly resist erosion by running water, and wide, shallow valleys and stream channels form characteristically in easily eroded materials, either bedrock materials or unconsolidated stream deposits in a broad floodplain.

The nature and amount of the suspended and bed loads of a stream at any point and at any particular time are a function not only of volume and velocity but also of the characteristics of the channel and bedrocks upstream. Clearly corrasion by a swift stream transporting a bedload of rock fragments and bowlders is greater than that by an equally swift stream carrying a heavy load of mainly very fine-grained clayey materials derived from soft rocks or alluvial floodplains in upper reaches of the stream. A stream charged with fine-grained sediment generally is not effective in downcutting and expends most of its available energy in sediment transport and lateral cutting. This conclusion is in accord with the observation that in many deep narrow valleys streams contain moderate suspended loads and bed loads, perhaps not moving at the present time, of a variety of rounded, resistant pebbles, cobbles, and bowlders.

In cross-section stream-cut valleys display a variety of profiles depending on the nature and geologic structure of the bedrock and the relative effectiveness of lateral erosion and downward cutting. In a broad characterization a distinction may be made between valleys which are *symmetrical* or *asymmetrical* in cross-section, as indicated in Fig.3-2 and 3-3.

In Fig. 3-2 downward cutting by a sinuous stream has produced a valley of symmetrical cross-section. The course of the stream presumably has remained essentially constant since the inception of downcutting, and widening of the valley is a consequence of gravitative movement of more or less weathered bedrock

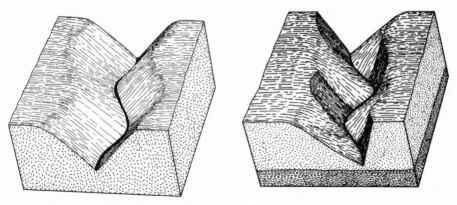

Fig.3-2. Stream-cut valley with symmetrical cross-section.

Fig.3-3. Stream-cut valley with asymmetrical cross-section.

materials on the slopes to the valley bottom where they were carried away by the stream.

In contrast, the stream valley in Fig.3-3 is asymmetrical because of a balance between downward and lateral erosion by the stream. Over-steepened slopes are formed on the outside curves in the stream by undercutting and are opposed on the other side of the valley by flatter slopes which result from gravitative movement of slope materials to the valley floor in the same manner as in symmetrical valleys. With time, the sinuosity of the stream tends to increase and the curves in the stream and the intervals where undercutting is prevalent, migrate downstream.

TERRACED VALLEYS

Many stream-cut valleys exhibit terraces, either in alluvium in the valley floor or in bedrock. Terraces are "paired" as in Fig.3-4 or "unpaired" as in Fig.3-5.

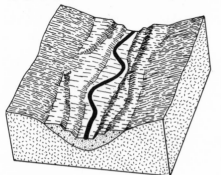 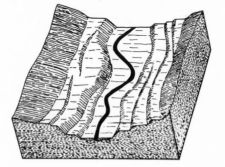

Fig.3-4. Paired terraces in alluvium.

Fig.3-5. Unpaired (meander) terraces in easily eroded shale.

Paired terraces indicate interruptions in downward cutting by the stream, whereas unpaired terraces (sometimes called "meander terraces") form when laterally shifting streams also are actively downcutting, thus leaving unpaired undercut banks on opposing sides of the valley. Unpaired terraces are not uncommon in valleys where, during periods of high water, streams actively remove large volumes of unconsolidated stream gravels or where bedrock (as in Fig.3-5) is easily eroded.

In the total history of a stream-cut valley terraces may form repeatedly, only to be later destroyed. Accordingly, the terraces that now exist may provide a record of only relatively recent erosional events in the valley. In cross-section terraces have step-like configurations, and, if the flat portions are considered to be remnants of former floodplains, where, for a time, lateral cutting by meandering streams dominated over downward cutting (a common interpretation), it becomes necessary to postulate a fortuitous sequence of events and processes that will result in intermittent development of two or more floodplains with decreasing elevations and decreasing widths.

A common explanation for paired terraces assumes either a rapid lowering of baselevel downstream (such as lowering of sea level) or uplift of the earth's crust. Either process will cause "stream rejuvenation" and will promote downward cutting below the previously existing valley floor. Multiple terraces require the assumption of repeated stream rejuvenations with successively smaller amounts of lateral erosion after each rejuvenation so as to preserve the step-like configuration of the terraces.

In many areas multiple terracing is more closely dependent on climatic variations than on changes in base level or uplift of the land surface. For example, in and near mountainous regions that were glaciated during the Ice Age, terraces can be correlated with repeated advance and recession of valley glaciers or ice caps of decreasing size.

Particularly in arid and semiarid regions stepped terraces may increase in lateral extent with time, and the total width of a flat surface at a particular terrace level is not the consequence only of lateral erosion by a stream meandering across a floodplain. The process effective in widening terraces is sheet-flood erosion and local gullying by the same process that results in development of extensive flat to gently sloping rock cut surfaces (pediments) at the bases of uplifted areas in arid and semiarid regions. Where the eroded subsurface materials are rich in clay extensive lateral transportation is promoted by the development of mud flows during periods of heavy rainfall.

The widening of terraces by sheet-flood erosion is indicated diagrammatically in Fig. 3-6. An assumption is made that at a particular time T_1 two terraces in shaly bedrock rise above a stream bed in alluvial gravel. At later times, T_2 and T_3 sheet erosion especially concentrated at the toes of the slopes defining each terrace have caused widening of the flat surfaces on each terrace and widening at the level of the

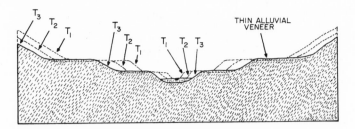

Fig.3-6. Widening of terraces by sheet flood erosion. T_1, T_2, T_3 indicate profile of valley at decreasing times.

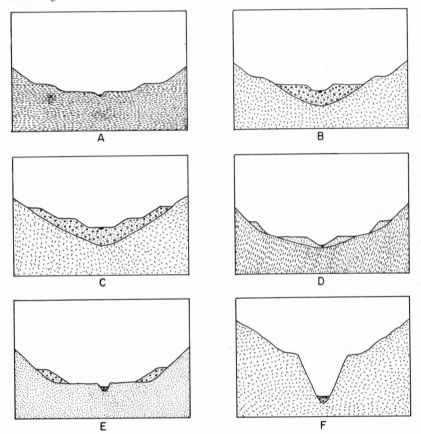

Fig.3-7. Terraced valleys.

A. Terraces in bedrock. Alluvium is scarce or practically absent.
B. Terraces in bedrock and alluvium.
C. Terraces in alluvium only.
D. Terraces in alluvium have been eroded to expose bedrock.
E. Remnants of terraces in alluvium preserved along sides of valley.
F. Terraces in rock in steep-walled valley.

stream. In effect, each flat surface acts as a base level that prevents downward erosion, but does not prevent erosion at the toe of the steep slopes between contiguous terraces or at the toe of the slopes at the outer edge of the highest terrace.

Several cross-sections of terraced valleys are shown in Fig.3-7 and indicate a few of the many kinds of interrelationships between terraces and the alluvial or bedrock materials in which they have developed. For safe design and construction of a dam and reservoir in a terraced valley, it is apparent that a knowledge of the origin, characteristics, and distribution of the materials in terraces, as well as the materials beneath the stream bed in the valley floor, is essential.

GLACIATED VALLEYS

As much as 10% of the world's land area is presently covered with ice in ice-sheets, ice-caps, and valley glaciers. A large proportion of this ice is concentrated in ice-sheets in Antarctica and Greenland, and the remainder is in ice-caps and valley glaciers in mountainous areas. During the Quaternary Era, which lasted for approximately 1.5 million years, ice covered at a maximum as much as 30% of the world's land areas. At least four, and probably five, advances and retreats have been documented. The last major advance reached its culmination about 20,000 years ago, and the glaciers that we see today are the residuals from retreat of the last advance.

Ice sheets of subcontinental dimensions moved over lowlands and mountainous areas alike and by scouring modified the shapes of landforms, and by deposition left huge volumes of glacial till and fluvioglacial till on the land's surface. Many areas buried by ice sheets now contain great numbers of lakes formed where ice gouged out bedrock or where glacial deposits dammed streams.

Of particular interest in the construction of dams and reservoirs are streamcut valleys that have been modified more or less extensively by glaciers moving through them. The ice in the valleys generally came either from ice caps which discharged masses of ice into the heads of valleys cut into a rolling or flat upland or from thick accumulations of snow which formed at the heads of valleys and were gradually converted into ice.

An idealized example of a glaciated valley in which the glacier originated in a basin, a *cirque,* quarried by ice action at the head of a valley is shown in Fig.3-8. Down the valley from the cirque erosion and quarrying have removed solid rock from the bottom and sides of a previously stream-cut valley and locally have excavated basins and steps in bedrock in the valley floor. Filling of the rock basins results in the development of so-called *tarn* lakes. In the lower reaches of the glacier a variety of deposits form at the sides, beneath, and at the terminus of the glacier.

Morainal ridges formed by deposition of glacial till along the sides of the glacier are called *lateral moraines,* and, when the terminus of the moraine remains stationary for a period of time, a *terminal* or *end moraine* is developed. Irregular masses of till erratically deposited by a retreating glacier are identified as *ground moraines.*

    Running water at the sides, within, and on the valley floor beneath the glacier, transports and locally deposits more or less well-sorted sands and gravels,

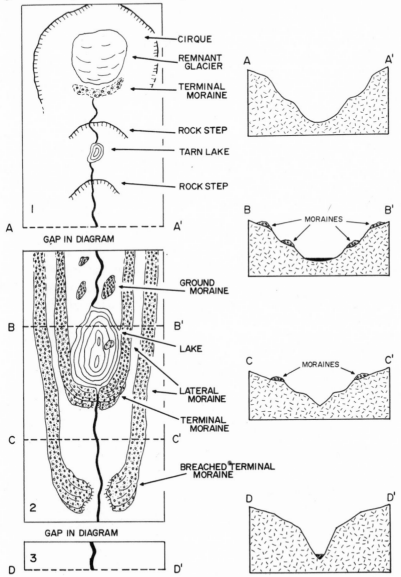

Fig.3-8. Idealized plan and sections of a stream and glacier eroded valley with two stages of glacier advance and retreat and prior and intervening periods of stream erosion.

and melt water flowing downstream from the terminus of the glacier transports and deposits *fluvioglacial deposits* for moderate to great distances.

In Fig.3-8 it is assumed that a period of stream erosion intervened between two stages of glacial advance and retreat and that ice in the second stage of glaciation did not fill the valley to the same extent nor did it extend as far down the valley as the ice in the first stage. As a consequence, a U-shaped valley has been excavated below the level of an earlier glaciated valley, also U-shaped in cross-section in the upper reaches of the valley, and glacial deposits were generated at two distinctly different times in lower reaches of the valley. Cross-sections at various locations in the valley also are shown in Fig.3-8 and indicate the changing valley configuration from the cirque downstream to the interval in the valley where deposition dominated over downward and lateral erosion and quarrying by the ice.

Section $D-D'$ in Fig.3-8 is a cross-section in the valley downstream from the valley interval occupied by ice and illustrates a typical, terraced, V-shaped, stream-eroded valley.

TRANSPORTED VALLEY FILL

We are concerned in this section with natural materials transported to a site of deposition in a valley from a more or less distant source by running water, glacial ice, or wind. The physical characteristics such as particle size and angularity of a deposit from running water at a particular location depend on the nature of the materials that become part of the stream load upstream, the transporting capacity of the stream, and the distance of transport. Reduction of *transporting capacity,* also called *competence,* is a consequence of decreased gradient, decreased volume, or damming of the channel.

Alluvial floodplains (Fig.3-9) are the most common result of deposition along

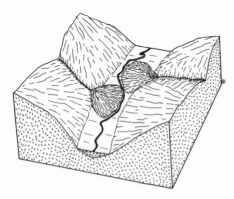

Fig.3-9. An alluvial floodplain buried in part by alluvial fans formed from tributary streams.

the course of a stream and consist of materials deposited in the stream channel and, by bank overflow, materials deposited outside the channel. A typical alluvial floodplain forms by alternating deposition and erosion by a sinuous, sometimes meandering stream which undergoes changes in location and configuration with time. The materials beneath the floodplain generally are more or less stratified and may exhibit extreme ranges in grain size, crude or well-developed stratification, cross-bedding, and cut-and-fill structures. Major erosion of bedrock beneath the floodplains in many major drainages is attributed to floodwaters associated with recession of glaciers during one or more glacial epochs, and the alluvial deposits, in some drainages several hundred feet thick, developed as transporting capacity diminished after the last glacial epoch.

Alluvial fans (Fig.3-9) are built up where tributary streams with high transporting capacity enter a valley occupied by a lower-gradient stream and accordingly a lower transporting capacity. Depending on the nature of source materials, fans consist of a fragmental material of a variety of sizes, ranging from bowlders to sand and silt or clay. Coarser materials at the apex of the fan tend to grade into finer substances near the margins, but with lateral shifts of the stream channel and fluctuating transporting capacity the vertical and lateral distribution of coarse and fine materials may become nearly random.

Deposition by ice produces *lateral moraines* along the sides of valleys, and/or *terminal moraines* at the terminus of a stagnating moraine, and *ground moraines* by irregular dumping of ice-enclosed till during ice retreat by melting. Lateral moraines commonly contain not only materials dragged along the edge of a moving ice body but, also, products of downslope movement of rocks and soils from exposures above the ice filling in the valley.

Typically, the materials in moraines, except where they have been reworked by water, are completely unsorted and, depending on the source of supply and the extent to which grain size has been reduced by corrasion, consist of particles and rock fragments ranging from clay size to bowlders many feet in diameter. An abundant ingredient of many till deposits is rock flour which is produced by mechanical breakdown by corrasion and crushing of quartz and feldspar in crystalline rocks.

In arid and semiarid regions or in regions close to the sea, valleys may be partly filled with shifting or more or less stabilized sand dunes. During the last ice age prevailing winds transported large amounts of dust derived from dry regions or from vast floodplains and deposited it in thin to thick accumulations of unstratified *loess* consisting predominantly of silt-size particles with minor amounts of fine sand and/or clay.

VALLEY FILL OF LOCAL DERIVATION

Unconsolidated deposits of local derivation on the valley sides and floors are described as *colluvial deposits* or *colluvium*, as contrasted with stream deposits, which are characterized as *alluvial deposits* or *alluvium*. Large colluvial deposits are said to be the products of *mass wasting* whereby large masses of earth material are moved downslope under the influence of gravity.

Weathering of rock rubble from mass wasting, especially in talus accumulations, causes a reduction in particle size, and generation of a soil cover. Accumulations on slopes of rock fragments mixed with products of weathering sometimes are identified by the term *taluvium* (mixed talus and colluvium). When this term is used, the definition of colluvium is restricted so as to include only the clayey to sandy end-products of weathering breakdown of coarser rock aggregates.

Mass wasting is a result of one or a combination of processes, including *sliding, flow,* and *heave.* Sliding in its purest manifestion occurs above a surface of shear. *Flow* can not be related to a shear surface, but, instead, shear is distributed through the moving mass and the movement resembles that of a liquid. *Heave* is characterized by expansion perpendicular to the surface of a layer of material, either by expansion by the addition of moisture or by expansion resulting from the freezing of interstitial water. Contraction produces a net dislocation downslope under the influence of gravity.

In consolidated materials on slopes, especially in bodies rich in clay-size particles, pore water introduced by seepage promotes downslope movement by exerting pore pressure and by tending to disperse fine-grained aggregates by attachment to particles by adsorption or, in swelling clays, by absorption. Pore pressure and particle dispersal work together to create instability by reducing shear strengths and, accordingly, easier slope failure under the influence of gravity. Slopes in wet materials generally are much flatter than slopes in dry materials of the same grain size and mineral composition.

Slope angles in unconsolidated surface deposits formed by mass wasting range from steep to flat, depending on particle size and effective pore pressures from seepage water. In aggregates containing large to small angular bowlders and numerous voids slope angels as high as 38° have been measured. Weathering reduction of such aggregates to form taluvium is accompanied by reduction of the slope angle to 25–28°. Lowest slope angles are found on coherent clayey materials, commonly about 10°, but less where clayey aggregates have spread out in mudflows.

The processes of mass wasting are intimately dependent on climate and weathering, topography as it is expressed in the area and angle of inclination of slopes, and on the nature of bedrock and the soil and vegetation covering it. Mass wasting is a consequence of slope failure, either slow or rapid, and is expedited by weathering which weakens bedrock by chemical composition or physical disintegration, or the presence of structural planes of weakness such as joints, faults, or bedding planes in bedrock and is retarded by an abundant cover of plant growth. Downslope movement as a consequence of transportation by small, ephemeral streams, rainsplash, and slope wash by sheets of water, is minimized by plant cover and is at a maximum in arid or cold regions where plant growth is sparse or absent.

In semiarid to humid regions where hot summers alternate with cold winters, microclimate plays an important role in inducing mass differential wasting within a single valley. At intermediate latitudes east—west valleys commonly are heavily forested on slopes inclined away from the sun and are sparsely covered with vegetation on slopes facing the sun. Forest cover retains ground moisture, promotes deep weathering, and reduces erosion by slope wash to a minimum, but rapid dessication, freezing and thawing, and expansion and contraction on slopes facing the sun, together with slope wash, inhibit plant growth and lead to mechanical breakdown and rapid downslope movement of the bedrocks. A common result is development of valley asymmetry because of different rates and kinds of slope reduction on opposing valley sides. In contrast, in humid, tropical and subtropical regions, where both sides of the valleys are generally covered with dense vegetation, valleys tend to be symmetrical and are capable of supporting relatively steep slopes.

Slow movement down slope of blankets of soils and fragmental rock accumulations resulting from chemical and/or mechanical weathering of bedrock is described as *solifluxion* and is an important process of slope reduction and valley filling. Depending on the competency of the stream in the valley to remove materials encroaching on the valley floor, thin or thick deposits may form. In environments where the soils develop by chemical and/or biochemical reduction of bedrocks the phenomenon, suggested in Fig.3-10, is widespread. Steady or seasonal

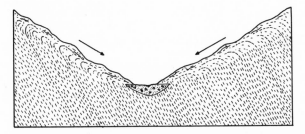

Fig.3-10. Solifluxion in a weathered layer above shale. Material moving downslope is covering an alluvial deposit of gravel.

down slope of clay-rich materials bends planar features within the weathered bedrock by differential movement, and at the surface tends to develop corrugations parallel to the contours. These corrugations have been called *terracetes*.

A spectacular kind of solifluxion is noted on valley slopes in cold regions with periglacial climates. In such regions, surface blankets rich in rock fragments produced by frost sundering and heave and usually containing alternately growing and waning permafrost layers or interstitial masses move seasonally toward the valley floor. Solifluxion on gentle slopes above a U-shaped glaciated valley is shown in Fig.3-11. Talus accumulations from the dislocated rock debris form at the base of the steep-walled glaciated valley.

Solifluxion (soil creep) primarily involves only a blanket of material covering a broad expanse of slope. More localized gravitative dislocation of soils and bedrock produces features that are identified as *landslides*. These may be slow or rapid and

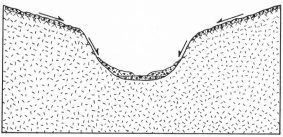

Fig.3-11. Solifluxion on gentle slopes under influence of a periglacial climate. Accumulations of talus have formed from solifluxion products at base of steep slopes in a glaciated valley.

TABLE 3-1

Classification of landslides

I.   *Falls in bedrock and soil.*
      A. By sudden detachment of bedrock and fall of rock from side of steep-walled valley.
         Rockfall
      B. By rapid detachment of soil from a steep exposure, usually in an undercut stream bank.
         Soilfall

II.   *Slides in bedrock and soils.*
      A. By dislocation of bedrock.
         1. Dislocation with rotation on a curved, concave surface.
            Slump
         2. Dislocation on an inclined, planar surface.
            Blockglide or *rockslide*
      B. By dislocation of soil.
         1. Rotational dislocation with rotation on a curved, concave surface.
            Block slump
         2. Dislocation on an inclined, planar surface.
            Block glide or *debris slide.*

III. *Flows in unconsolidated materials.*
      A. Dry flow.
            Rock fragment flow, sand run, loess flow.
      B. Flow of moderately wet or water-saturated materials.
            Earthflow, debris avalanche, mudflow, sand or silt flow, debris flow.

IV.   Complex flows, combining two or more of above mechanisms.

---

may involve small to very large volumes of materials. A classification of landslides useful in engineering practice, adapted from a classification by Varnes (1958), is given in Table 3-1.

Examples of mass wasting in valleys are illustrated in Fig.3-12–3-14.

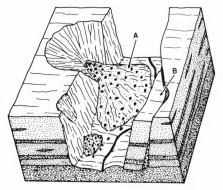

Fig.3-12. Talus cones (A) formed by rockfalls in steep-walled canyon in sandstone. On the right (B) a large block has become detached by settlement in a shale layer.

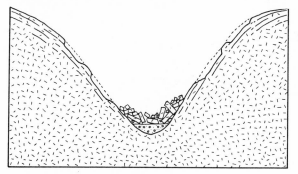

Fig.3-13. Failure of steep slopes along fractures parallel to surface and produced mainly by elastic rebound as a consequence of unloading by stream erosion.

In Fig.3-12 a canyon with nearly vertical sides has been eroded in sandstone with shale layers. Slope failure producing rock falls has caused the accumulation of talus cones, and settling of a block in a shale substratum has caused its dislocation along a steep surface. Features illustrated in Fig.3-12 are common in semiarid and arid regions underlain by thick massive sandstone layers or by horizontal or nearly horizontal layers of volcanic rocks.

Rapid downward cutting of valleys by streams and removal of ice in glaciated valleys create unbalanced forces in bedrock, especially in monolithic, relatively unfractured crystalline rocks such as granite. The removal of a pre-existing load, whether it is the rock excavated by a stream or by ice that disappears by melting, promotes *elastic rebound.* This phenomenon, probably assisted by long-term changes in rock temperature with changing weather or climate and, to a minor extent, by chemical weathering results in the development of joints parallel and subparallel to the surfaces of rock exposures (Fig. 3-13). Rock slabs breaking away from the joints commonly reveal an onion-like succession of joints and the end product of repeated separation of slabs frequently is the formation of conspicuously rounded knobs of rock and accumulation of rock debris in the valley floor.

In cold, deeply dissected regions that have been glaciated and are now characterized by a periglacial climate with severe winters and short summers, slope failure promoted by ice wedging and gravitative collapse causes reduction of over-steepened slopes. Several kinds of valley-filling processes in glaciated regions are shown in Fig. 3-14, where slopes in both bedrock and unconsolidated glacial till

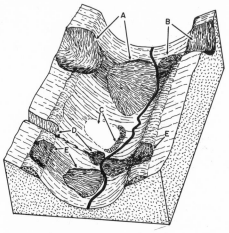

Fig. 3-14. Glaciated valley partly filled by landslides and alluvial materials. A = talus cone from slope failure in bedrock; B = alluvial fan deposited from a stream occupying a glaciated hanging valley; C = protalus rampart formed at base of a seasonally fluctuating snowbank; D = alluvial fan from tributary stream-cut valley; E = talus accumulations from slope failure in unconsolidated glacial till in a lateral moraine.

have failed. In addition, Fig. 3-14 shows alluvial fans deposited from streams issuing from hanging valleys tributary to the main valley.

A not uncommon feature of valleys in recently glaciated areas is shown in Fig. 3-15. Melting of glacial ice in a cirque and in lower reaches of a steep-walled valley excavated by ice has generated unstable, oversteepened slopes which by gradual breakdown into rock fragments by ice wedging and gravitative collapse have produced a series of coalescing talus cones. The accumulation of rock fragments, together with pulverized rock particles in a matrix, move down the valley as a *rock stream* or *rock glacier*. Presently active rock glaciers are known to contain an under-stratum of clear ice or cells of dirt-laden ice in at least the lower half of the moving accumulation, and it is probable that seasonal melting and refreezing of ice is the cause of gradual, but measurable flow of rock debris down the valley floor.

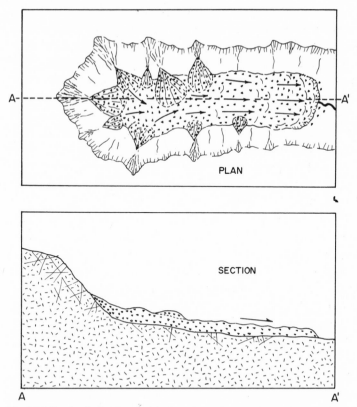

Fig.3-15. Rock glacier in plan and section. Note coalescing talus cones at upper portion of valley. Arrows indicate direction of movement of rock debris down valley floor.

SLOPE STABILITY AND SLOPE FAILURE IN VALLEYS

Downslope movement of materials on and beneath slopes of valleys under the influence of gravity involves a wide spectrum of processes. At one extreme is dislocation of individual particles by rain-splash or wind, and at the other is movement of large volumes of rock or unconsolidated materials along curved or planar surfaces of shear inclined toward the valley floor.

In general there are two types of surfaces of rupture along which masses of soil or bedrock are displaced toward a valley floor. In materials of low strength, such as unconsolidated deposits, some shales and mudstones, and initially strong rocks weakened by fracturing and/or weathering, mass movement occurs mainly along curved shear surfaces. In brittle elastic rocks of moderate to high strength, such as quartzites, crystalline limestone, and most igneous rocks, shear surfaces and fractures developed by tension are essentially planar.

The reasons why curved surfaces of rupture develop in weak materials and planar surfaces are generated in strong elastic substances are not fully understood. However, if a spoon-shaped surface in weak material is compared with a spherical surface, it can be pointed out that the volume (and mass) of material above the surface of shear is at a maximum relative to the area of the surface. Elastically-behaving natural substances generally have notably higher densities than weak materials and are capable of transmitting directed forces by elastic strain to a much greater extent than weak substances. Under the circumstances, the most efficient mechanism of shearing, one involving a minimum critical mass (and volume) and a minimum area of surface of rupture, is failure along a planar surface.

Analysis of the stability of slopes leads to many perplexing problems because of the fact that failure occurs only when some critical mass is reached owing to accumulation of body forces. Analysis of the effects of pressures on foundations from dams and bodies of water in reservoirs can be made with relative ease mathematically or by the use of vectors which have directions and magnitudes and which, after taking into account shear strengths, enable reasonably accurate prediction of the locations of surfaces of rupture. In contrast, prediction of the locations and attitudes of rupture surfaces associated with slope failure in terms of body forces generally is highly unsatisfactory because of lack of data and insufficiently comprehensive theory. Recognition of instability in slopes, then, must to a large extent be based on empiricism and experience derived from observation of many slope failures, and prediction of slope failure commonly is made on the basis of rule of thumb analysis.

Forces which may act on a body of matter are *surface forces* and *body forces*. Surface forces act on an external surface of a body and come about from physical contact of the body with another one. Examples are the forces exerted by dams on foundations and by pressure apparatus used in testing strengths of materials in the laboratory. These are forces which can be measured or calculated and translated into stresses. Body forces act throughout a mass of material and do not require physical contact with other bodies. Examples are gravitational, magnetic, and inertial forces.

A vast amount of information is at hand concerning the nature of the interaction of surface forces with bodies of matter. Much less is known about body forces, although experiments with centrifuges in which centrifugal forces are created have given some insight into the strains caused by these forces.

The distinction between body forces and surface forces is brought out by comparing definitions of measures of their intensity. Intensity of surface forces is measured by *stress*, that is, *force per unit area* of application of the force. In contrast, the intensity of a body force is expressed as the *force per unit volume*. The body force of concern in the present context is, of course, the force of gravity.

SHEAR FAILURE IN WEAK MATERIALS

Natural materials that are weak and tend to shear along curved surfaces commonly have sufficient cohesive strength to stand in slopes of about 10–35°, if not saturated with water. Water, by penetrating slide-prone materials in slopes, greatly reduces their strength by adding weight and by causing increases in pore pressures which promote loss of cohesion. The effects of penetration of various kinds of materials in slopes by water are discussed in Chapter 5 and 6 and will not be reviewed here.

As a practical matter, prediction of the location in an actual slope of a curved surface of shear and the volume of material above the surface is generally very difficult, if not impossible. However, direct observation of slopes in the field and testing of the materials in them enables identification of general conditions that are likely to promote slope failure, although the exact locations and magnitudes of the dislocations can not be precisely ascertained. Many unstable natural slopes show evidences of failure in the recent or distant past, and the certainty exists that, if the

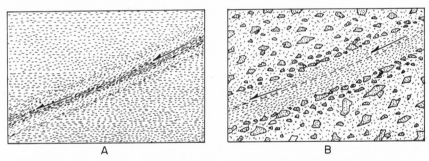

Fig.3-16. Reorientation and segregation of inequidimensional particles by shear dislocation.
A. Clay minerals of micaceous habit in a shale are rotated parallel to plane of shear.
B. Clay minerals in a mixture with rock fragments are concentrated in the vicinity of a plane of shear while rock fragments are eliminated from zone of active shearing.

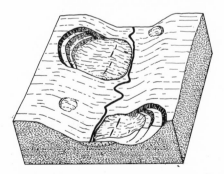

Fig.3-17. Slump in shale above a spoon-shaped surface of shear. Material from slump buries gravels in an alluvial floodplain causing changes in the course of the stream.

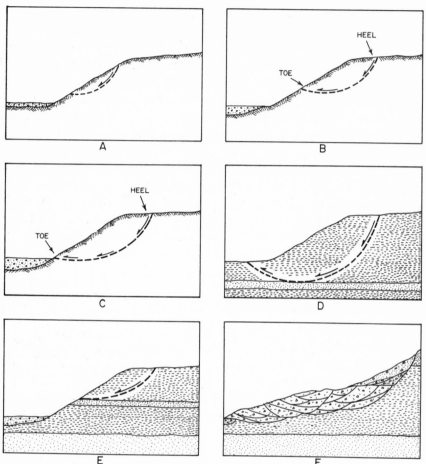

Fig.3-18. Cross-sections showing locations of potential surfaces of shear in weak materials in slopes.

A, B, C. Failures involving various intervals of slopes.

D. Base failure in shale limited at depth by a layer of sandstone.

E. Slope failure in shale above a sandstone layer.

F. Repeated slope failure caused by loading of heel of slide by rock and soil debris from steep slopes above.

environment that promoted failure in the past is recreated, slope failure will be reactivated anew.

Initiation of slope failure requires that gravitative body forces acting on a mass produce internal stresses that result in shear failure along a surface with a location determined by the geometry of the slope and the shear strength of the material. As a consequence of failure, stresses are released to a certain extent by dislocation of the mass above the shear surface, but frictional forces remain along the surface and tend to oppose sliding. Once sliding has been started, continued or

intermittent movement over the shear surface can be expected, and, in critical
situations, every effort should be made to forestall the beginnings of slope failure
by taking appropriate remedial measures.

The reason why a mass of material involved in slope failure continues to be
steadily to intermittently active over a short to long interval of time is suggested in
Fig.3-16. When the shear strength of the material in a slope has been exceeded,
movement along the shear surface and in its immediate vicinity tends not only to
reduce the grain size of particles shearing past each other, but, if the particles are
inequidimensional, also tends to rotate them so that their long axes are parallel to
the shear surface. In materials of contrasting grain sizes shearing causes not only
reorientation, but also segregation of smaller particles in a shear zone. Larger
particles are rotated and forced out of the aggregate. The net result is development
along the shear surface of a zone that continues to be less resistant to shear
dislocation than original undisturbed materials above and below the surface.

In three dimensions many slumped masses in weak materials are bounded
below by a spoon-shaped rupture surface, as in Fig.3-17. Cross-sections showing
various locations of curved shear surfaces in slopes on weak materials are illustrated
in Fig.3-18. Fig.3-19 shows examples of accumulations of unconsolidated materials

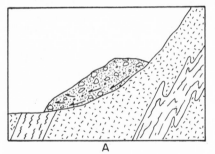 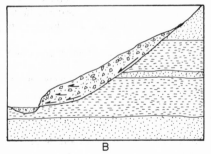

A	B

Fig.3-19. Accumulations of slide-prone, unconsolidated materials on inclined bedrock surfaces.
A.,Glacial till on floor and side of a glaciated valley in granite and gneiss.
B. Talus accumulation resting on an inclined bedrock slope in shale and sandstone.

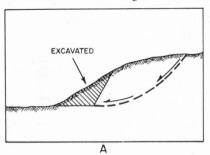 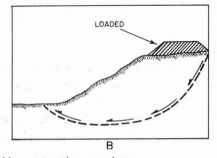

A	B

Fig.3-20. Slope failure in weak materials caused by construction procedures.
A. Slope failure is caused by excavation of the toe of a previously stable slope.
B. Slope failure is promoted by loading by construction of a road embankment.

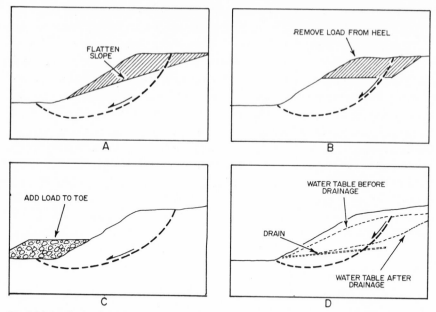

Fig.3-21. Methods of stabilizing slopes in weak materials that have failed or are likely to fail.
A. Stability is increased by excavating a flatter slope.
B. The tendency to fail is reduced by excavating material from the heel of an actual or potential slump.
C. Movement along the curved shear surface is prevented by loading the toe of an actual or potential slide.
D. Stability is increased by draining water from the slumped material.

that rest on inclined bedrock surfaces and are susceptible to shear dislocation both along the bedrock surfaces and within the interiors of the accumulations.

In reservoir sites or in the location of a dam slumps along curved shear surfaces in slopes that are relatively stable if left undisturbed commonly are initiated by construction procedures and may result in unwanted downslope displacements of small to large volumes of earth materials into the reservoir or the excavation for the dam. Examples of artificially induced slope failure are indicated in Fig.3-20.

Alternative methods for slope stabilization in weak materials are indicated in Fig.3-21. Any of the methods except that indicated in Fig.3-21D has possible utilization in stabilizing slopes in reservoirs either before or after slope failure. The method indicated in Fig.3-21D requires installation of drains to lower the groundwater table in the slide material and, obviously, has no application in the slopes below high water level of a filled reservoir. However, drainage may serve a useful purpose in decreasing the hazard of slope failure in road excavations and embankments and in slopes in weak materials above high water elevation in reservoirs.

GRAVITY-SLIP DISLOCATIONS IN STEEP-WALLED VALLEYS

The alluvial, glacial, and landslide deposits on the floors and sides of valleys generally have locations, configurations, and physical properties that are identified in the field with relative ease. During planning, design, and construction of a dam and reservoir an assessment of these deposits can be made without difficulty, and appropriate measures can be taken for their removal or stabilization.

In many steep-walled valleys, stream-cut or glaciated, a relatively inconspicuous kind of slope failure is present, especially in highly competent, crystalline igneous and metamorphic bedrocks. Although they may not be easily observed, gravity-slip surfaces may be present in bedrock as indicated in Fig. 3-22 and contribute to the instability of the foundation and abutments of a dam that might be constructed at the site.

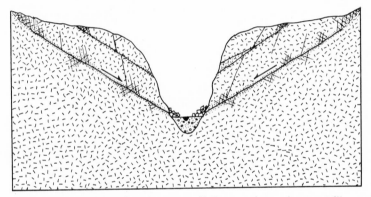

Fig.3-22. Gravity-slip surfaces in steep-walled canyon in massive, crystalline rocks.

The slip surfaces commonly are inclined toward the valley bottom at angles of 25–35° from the horizontal and where exposed commonly reveal slickensides and narrow, discontinuous lenses of crushed rock. Movement along the surface generally is only a few inches or a few feet, but sufficient fracturing above and below the slip surface locally is generated to enable free flow of groundwater and, under some conditions weathering of the rock and formation of springs where the slip surface intersects the valley bottom.

The slip surfaces appear to be a consequence of downcutting by the stream to a depth such that the dead weight of the rocks in oversteepened slopes becomes sufficient to cause shear failure and release of accumulated stresses. Elastic rebound by unloading during stream excavation may be a contributing factor in the rupture of the rocks.

Gravity-slips faults with angles of dip toward the valley floor in excess of 35–40° produce down-dip block movements and mass collapse of slopes. Frictional

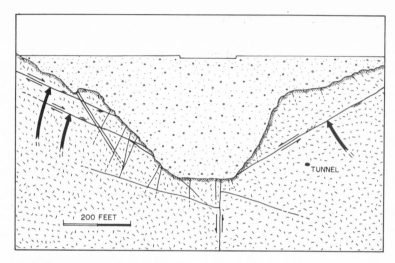

Fig.3-23. Faults in granitic foundation rocks at Gross Dam, Colorado. Gravity-slip faults are indicated by arrows. Other faults were present in bedrock before the gravity-slip faults developed, and are geologically much older.

Fig.3-24. Gravity-slip surfaces (indicated by arrows) in the left abutment of the keyway excavation for Gross Dam, Colorado. (Photo by Denver Board of Water Commissioners.)

resistance along the surfaces of rupture generally is not adequate to restrain movement by sliding to the same extent as over surfaces of lower inclination.

The kind of slope failure indicated in Fig.3-22 is different from failure in steep-walled canyons in horizontally layered rocks where detachment of blocks on canyon walls occurs along nearly vertical fractures.

An actual example of gravity-slip faults in massive granite is illustrated in Fig.3-23, which shows faults of a variety of orientations intersecting bedrock at Gross Dam in Colorado. The gravity-slip faults are indicated by arrows. Movement along the faults was of the order of only a few inches to a few feet but was sufficient to produce striations and mullion structures. Circulation of groundwater along the faults produced localized weathering alteration of the granite.

Fig.3-24 is a photograph of the left (north) abutment of the excavation for the keyway for Gross Dam showing gravity-slip surfaces dipping toward the canyon floor.

OBSTRUCTIONS IN STREAM VALLEYS

Partial to complete obstruction of stream flow by a natural process or construction of a dam and impoundment of water in a natural or artificial reservoir creates a dramatic change in stream regimen. Two expected consequences are deposition of all or much of the suspended and traction load transported by the stream into the reservoir and a more-or-less spectacular increase in downward and/or lateral erosion by clear or desilted water downstream from the obstruction.

Reservoir-wide siltation by clay- and silt-size particles and encroachment by coarser-grained materials in deltaic deposits are a major threat to the continued, useful life of a reservoir, especially of a reservoir supplied by a stream transporting large amounts of particulate matter. Reduction by artificial means of the normal average load of a stream entering a reservoir is extremely difficult, particularly if the stream flows out of a large drainage basin containing a complex variety of rock and soil types.

Attempts to control the rate of filling of reservoirs by sediment may include construction of dams and reservoirs to intercept sediment upstream from a major facility, such as a large dam for electric power generation, and regional programs for soil stabilization and conservation in upstream drainage basins.

LAKES AND LAKE BASINS IN VALLEYS

Many valleys contain lakes of diverse origins. Not uncommonly dams are constructed to enlarge the storage capacity of an existing lake or to take advantage of a basin formerly occupied by a lake which no longer exists because of alluvial filling and/or breaching of the natural dam that impounded the lake.

Lakes are numerous in many areas of the earth, especially in areas that were covered by ice caps during the Ice Ages and in areas where basins have formed by dissolution and collapse of limestone caverns (karst regions). However, our primary concern here is with lakes that are present or have existed in the past in valleys. The many processes that form lakes in valleys are summarized in Table 3-2. In general,

TABLE 3-2

Lakes in valleys

I.	Formed behind dams resulting from mass wasting (i.e., landslides).
II.	Formed by glacial action. A. Lakes in basins excavated from bedrock by ice action. *Tarn lakes* B. Lakes in basins within or behind deposits of glacial till, especially terminal and lateral moraines.
III.	From stream action. A. Floodplain lakes in abandoned river channels. *Oxbow lakes* B. Lakes behind alluvial fans from tributary streams.
IV.	From volcanic activity, especially by damming of streams by lava flows.
V.	From crustal dislocations, especially faulting.
VI.	From organic processes, as, for example, *beaver ponds* behind dams.
VII.	In basins from collapse of limestone solution caverns in karst areas.

valley lakes are most numerous in glaciated valleys and on broad alluvial flood-plains, where oxbow lakes occupy abandoned stream channels developed during stream meandering.

The ultimate fate of a lake is its extinction, either by filling of the lake basin or failure or erosional removal of the natural barrier impounding the lake. Filling is a more or less complex process involving influx of alluvial materials, accumulation of organic materials, and local downslope movement of materials of solifluxion and mass wasting. The internal structure of the materials filling a lake basin may range from simple to complex, depending on the kinds and intensities of the processes that act to fill the basin and is of concern in engineering planning and design of a dam and reservoir at the site of a filled basin.

Filling of a lake by a stream is indicated diagrammatically in Fig.3-25. The rate of filling and the relative properties of deposited materials in various size ranges

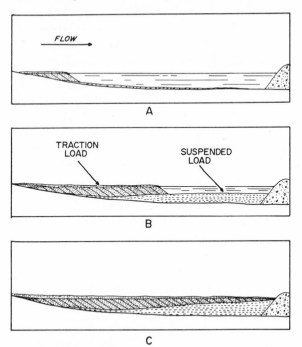

Fig. 3-25. Stages in filling of a lake basin by unconsolidated sediments transported into lake by a stream. See text discussion.

depends on the volume of the stream, its transporting capacity, and the nature of the materials available for erosion and transportation in upper reaches of the stream valley. In Fig.3-25 it is assumed that the load of the stream ranges in size from coarse gravel to clay size, and that part of the load is transported by traction and part in suspension.

Deposition of the traction load builds up a gravel delta where the stream enters the lake and produces inclined *foreset beds* or *layers,* and, on top of these, relatively flat *topset beds.* Suspended material of silt and clay size is carried out into the lake where it eventually settles out to generate *bottomset* beds in which particle size decreases toward the stream exit of the lake. Seasonal variations in the amount and character of the stream load produce alternating very fine-grained layers and somewhat coarser-grained layers in the bottomset beds. The layers are sometimes identified as *varves.* Typically, coarser-grained, thicker layers are deposited in spring and summer and thinner, very fine-grained layers in fall and winter. Gradual encroachment by the foreset and topset beds in the delta eventually results in complete burial of the bottomset beds and filling of the lake basin.

In valleys with oversteepened slopes mass wasting contributes to the filling of

82   EXCAVATION AND FILLING OF VALLEYS

a lake basin and, with materials carried in by streams, produces a complex pattern
of layers and lenses of unconsolidated aggregates with widely different particle sizes
and permeabilities. In addition, wave action and ice shoving erode, rework, and
transport materials around the shore line of the lake forming lake terraces and
undercut banks. Erosion along the shoreline may promote slope instability and
commonly generates deposits of sand and gravel offshore in the lake basin.

The stages in the filling of a lake in a glaciated valley are indicated diagrammatically in Fig.3-26 and 3-27. The processes are of the same kinds that might
operate in a lake in a steep-walled stream-cut valley. Fig.3-26 shows in plan a lake
that has formed behind a morainal dam and that has been filled gradually by
sediments from a main stream and a tributary stream and by-products of mass
wasting from intermittent slope failure. Fig.3-27 indicates gradual filling of the
basin in a typical cross-section near the lower end of the lake. No attempt is made
in the cross-section to indicate bedding and variations in grain size of the materials
of stream derivation. Progressive filling of the lake has produced an intertonguing
relationship between more or less permeable deposits of alluvial origin and moderately to highly permeable products of mass wasting.

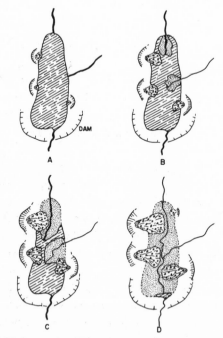

Fig.3-26. Progressive filling of a glacial lake basin by alluvial deposits and products of mass
wasting. Material transported into the lake by streams is indicated by stippling.

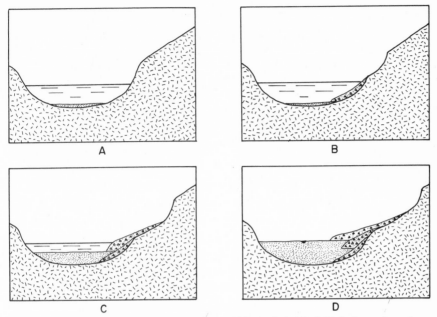

Fig.3-27. Typical cross-section near lower end of lake basin in Fig.3-26 during various stages of filling. Deposits of stream origin are indicated by stippling.

REFERENCES

Carson, M. A. and Kirkby, M. J., 1972. *Hillside Form and Process.* Cambridge Univ. Press, London, 475pp.

Morisawa, M., 1968. *Streams, their Dynamics and Morphology.* McGraw-Hill, New York, N.Y., 175pp.

Varnes, D. J., 1958. Landslide types and processes. In: *Landslides and Engineering Practice.* Highway Research Board, Spec. Rep., 29, Washington, D.C., pp.20–47.

Yatsu, E., 1966. *Rock Control in Geomorphology.* Sozosha, Tokyo, 135pp.

Chapter 4

VALLEYS IN ANISOTROPIC ROCKS

INTRODUCTION

In Chapter 3 processes contributing to the development and filling of valleys in essentially homogeneous bedrocks were reviewed. No attempt was made to relate the locations and characteristics of valleys to subsurface geologic structures. However, as streams (and glaciers) excavate or modify valleys, they commonly encounter a variety of bedrock materials with physical properties and space distributions that, during the history of a stream, may exert a profound influence on the configuration of an individual valley in plan and cross-section and on the pattern of development of a network of streams in a drainage basin.

The relationship of a valley and the stream in it to geologic structure in bedrock may be described as *accordant* or *discordant,* depending on whether the location of the valley conforms closely to the disposition in space of bedrocks of different physical properties and erodabilities or the valley cuts across geologic structures with little regard for lateral and vertical differences in physical properties of bedrocks. Many valleys locally display both accordant and discordant features along their lengths.

The initiation and evolution of a single valley and/or the valleys of a drainage system depend mainly on two factors: (1) the nature of the surface over which the streams first begin to flow, and (2) the geologic structure of the bedrock that is encountered as streams excavate their valleys. The topographic relief of the "initial surface" may be slight or extreme, but, in any event, as a result of geological processes such as uplift of a gently inclined flat surface or large to small scale dislocations of bedrock by faulting and/or folding, streams are generated which follow lines of steepest gradient on the land surface. Streams whose courses are determined by the topography of the "initial surface" are called *consequent* streams and excavate comparatively straight valleys in gently inclined, nearly planar surfaces, and valleys of diverse, sometimes circuitous, courses in complexly dislocated terranes.

If a consequent stream is able to maintain its course by downward cutting during a slow differential uplift across its trend it is said to be an *antecedent stream.* Contrasted with an antecedent stream is a *superimposed stream* which maintains its course as the valley is incised, although changing geological conditions are encountered. For example, a consequent stream excavating a valley in flat-lying

sedimentary rocks might encounter entirely different geological conditions below
an unconformity. If the stream maintains its course without regard for the changed
environment, it becomes a superimposed stream in discordance with the structural
features of the bedrock.

Antecedent and superimposed streams generally are trunk streams in major
drainage systems. Once a consequent drainage system has become established on an
"initial surface" *subsequent tributary streams* develop and grow headward by
erosion along zones of weakness in bedrock. Because subsequent streams are
accordant and in harmony with bedrock structure the patterns developed by them
reveal much concerning geological conditions below the earth's surface.

Interpretation of the history of a stream and the valley it has eroded is vital
to an understanding of the structural conditions in bedrock and the stability of
slopes in a valley interval where it is proposed to construct a dam and reservoir.

Special features associated with erosion and dissolution of highly soluble
rocks including limestone, dolomite, and gypsum are not discussed in this chapter,
but, instead, are given special treatment in Chapter 7.

ANISOTROPISM IN ROCKS

Most rocks are anisotropic, that is, they display different properties in
different directions. In contrast, isotropic rocks have the same properties in all
directions. A distinction is made between *primary* or *original anisotropism,* and
secondary anisotropism. Primary anisotropism is a consequence of the processes
that were active during the formation of a rock body and is seen in planar and
linear growth fabrics in igneous and metamorphic rocks and stratification and other
depositional features in sedimentary rocks. *Relict primary anisotropism* is inherited
from a previous condition of a rock, and is observed, for example, in layered
metamorphic rocks formed by recrystallization at high pressures and temperatures
of stratified sedimentary rocks. Secondary anisotropism is superimposed on rocks
by a variety of processes that include development of faults, joints, and folds, and
directional alteration or dissolution by surface waters and/or groundwater solutions. Secondary anisotropism may be imposed either on rocks with original
anisotropism or on isotropic rocks.

In following sections of this chapter attention is given to the controls exerted
by anisotropism in bedrocks on the development of characteristic patterns in
drainage systems and large and small-scale features in individual valleys.

VALLEYS IN FLAT-LYING LAYERED ROCKS

In the present context "flat-lying" refers to the attitude of layers bounded above and below by laterally extensive parallel or subparallel surfaces of discontinuity with the understanding that internal elements of fabric such as cross-beds may be inclined at small to large angles to the major surfaces of discontinuity.

In nature the great preponderance of flat-lying layers are generated by deposition on surfaces of low inclination of materials transported by wind, water, or ice. Materials comprising the layers display a broad spectrum of physical and chemical properties depending on their origin and their histories subsequent to deposition. Flat-lying layers also are generated by volcanic activity which causes lava flows and pyroclastic deposits to spread out over the earth's surface, by intrusion of molten magma between layers of flat-lying rocks to form sills, and by weathering of bedrock below a surface of low relief. The last process results in some localities in the generation of residual layers that are rocklike and highly resistant to erosion. Upward and downward migration of dissolved materials may produce layering resembling that developed by processes of sedimentation. Examples are seen in ferruginous laterite (ferricrete) in seasonal tropical areas and siliceous accumulations (silcrete) in some dry areas. A general term for resistant capping layers formed by weathering is *duricrust*.

Many valleys in flat-lying rocks are essentially symmetrical in cross-section. However, differences in microclimates on opposing sides of deep valleys, especially in east—west valleys, may lead to notable asymmetry by promoting different rates and kinds of slope reduction by weathering. A symmetrical valley in horizontal sedimentary rocks is shown in Fig.4-1. The step-like cross-section is a consequence

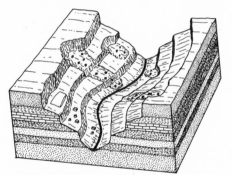

Fig.4-1. Symmetrical valley excavated in horizontal sedimentary layers of different compositions and physical properties.

of different slope angles and modes of slope failure in a succession including sandstones, shales, and a limestone layer. Vertical faces produced by steep slab dislocation along widely spaced joints in the sandstone layers are typical. Valley profiles in horizontal layered successions, some of which lie above unconformities, are illustrated in Fig.4-2.

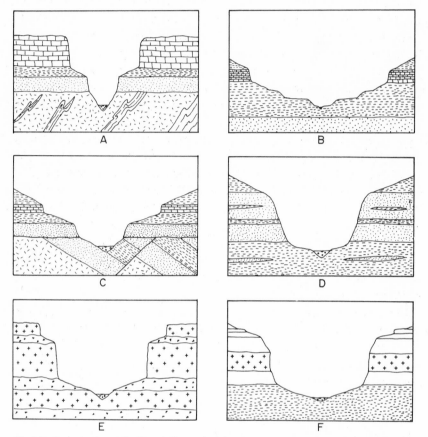

Fig.4-2. Cross-sections of valleys in layered rocks.
A. The stream has penetrated crystalline igneous and metamorphic rocks below an unconformity.
B. A limestone layer supports steep slopes and a broad, terraced valley has developed in shaly rocks.
C. The stream has cut through an angular unconformity and has moved laterally by erosion in a shale layer.
D. Glaciated valley in horizontal shale and sandstone layer.
E. Steep slopes have developed in lava flows and flatter slopes in pyroclastic deposits.
F. Vertical slopes develop in an igneous sill intruded along a sedimentary bedding plane.

Spectacular, large-scale examples of stream erosion in flat-lying layered rocks are observed along several intervals of the Colorado River in the western United States (Fig.4-3 and 4-4). Precipitous slopes have developed in resistant layers by falling away of small and large slabs of rock bounded by steeply-dipping to vertical fractures, and gentler slopes have formed over softer, less resistant layers. The

Fig.4-3. Glen Canyon Dam on the Colorado River in Arizona. Canyon has been cut in a massive to cross-bedded, thick sandstone layer. (Photo courtesy of the U.S. Bureau of Reclamation.)

Fig.4-4. Aerial view of Marble Canyon, Arizona, looking down on the Colorado River. (Photo courtesy of the U.S. Bureau of Reclamation.)

fractures in some locations are dislocations along faults but elsewhere have been generated parallel to the canyon walls by elastic rebound, uneven settlement of moderate to high strength layers into underlying weaker layers, and by collapse resulting from undercutting of the canyon walls.

VALLEYS IN FOLDED ROCKS

Valleys in folded layered rocks are *accordant* or *discordant* with respect to the structures in bedrock. Accordant valleys are located in the troughs of regionally extensive synclines developed by earth flexures and were eroded by consequent streams whose courses were determined by topographic lows which coincided with structural lows in bedrock. If the folds are symmetrical, the drainage patterns of tributary streams also tend to be symmetrical, as in Fig.4-5.

Folds in rocks are produced by a complex process of dislocation involving bending, shearing, or slipping on a large to small scale, and, in some rocks, partial to complete recrystallization. *Anticlines* are upfolds, and *synclines* are downfolds. A

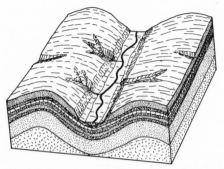

Fig. 4-5. Consequent drainage pattern developed in trough of a symmetrical syncline.

line joining the highest points of an anticline as defined by a particular bed is the *crest* of an anticline, and the *crest plane* is the plane passing through the crests of a succession of layers in the anticline. Similarly, we may define the *trough* and *trough plane* of a syncline. The *axial plane* is an imaginary surface that approximately bisects a fold and may be a plane or curved surface. A line in a folded layer which passes through points where the dip changes most rapidly is identified as the *hinge* or *hinge line* of the fold. Hinges may be horizontal or inclined, and when crests, troughs, or hinges are inclined from the horizontal, the folds are said to be *plunging.* A *dome* is an upfold which dips away in all directions from a high point.

Discordant valleys in folded terranes trend across folds, that is, they intersect the crests, troughs, and hinges of the folded rocks, as opposed to accordant valleys which follow troughs of synclines. Discordant valleys usually are eroded by initially consequent streams which maintained their courses in spite of changing rock conditions with depth and thus become superimposed streams. Less frequently, streams continue to cut downward without change of location through folds which form across their paths (antecedent streams). Spectacular topographic effects at times result from downward and lateral erosion by streams which encounter different kinds of rocks in successions of folded layers.

Valleys in simply folded rocks in monoclines, as in Fig.4-6—4-8, commonly

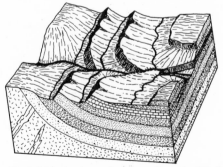

Fig.4-6. Hogbacks and strike valleys in tilted, eroded sedimentary rocks. Mesas have formed where the major stream has cut a deep valley below horizontally-layered rocks.

Fig. 4-7. Warner Valley, Utah. Note shallow valleys between hogbacks held up by tilted, resistant sedimentary layers. (Photo courtesy of U.S. Bureau of Reclamation.)

exhibit *ladder* or *trellis* drainage patterns in which valleys in soft layers *(strike valleys)* alternate with parallel to subparallel *hogbacks* or *cuestas* held up by harder layers. In trellis drainage a major stream superimposed on the structure has maintained its course while tributary subsequent streams excavated valleys in the soft layers by headward and downward erosion.

In many areas where trellis (or ladder) drainages have developed, *stream capture* operates to divert the flow of some of the streams. A consequence at times is formation of valleys which contain *misfit streams,* that is, small or ephemeral streams which now occupy valleys apparently eroded at some time in the past by much larger streams, and *windgaps,* stream-cut notches in hogbacks. The sequence of events that lead to stream capture in a trellis drainage in a hogback area is indicated in Fig.4-9. In Fig.4-9 it is imagined that two major parallel streams with somewhat different volumes of flow cross a succession of hard and soft layers. Headward and downward erosion by tributary streams in soft layers eventually results in capture and diversion of the smaller of the major streams by the other, thus leaving an essentially abandoned stream valley donw-valley from where the smaller stream has been diverted.

Fig. 4-10 indicates a hogback developed by downward and lateral erosion in plunging folds containing layered rocks with differing degrees of resistance to erosion. The major streams to which the smaller streams in Fig. 4-10 are tributary are not shown.

Fig.4-8. Flaming Gorge Dam, and power plant, Utah, constructed in a deep gorge in inclined sedimentary rocks. Note hogbacks in distance. (Photo courtesy of U.S. Bureau of Reclamation.)

Intrusions of magma that develop small to large igneous stocks or laccoliths and upward movement of salt in evaporite successions may result in formation of *domes* in layered rocks, as in Fig.4-11. Erosion of a dome by streams produces a characteristic pattern of major and tributary streams. Where sedimentary layers of differing resistance to erosion are encountered circular or elliptical strike valleys alternating with hogbacks are generated.

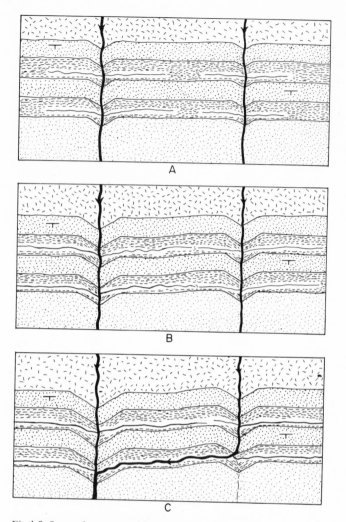

Fig.4-9. Successive stages of headward advance of tributary streams along soft layers to cause one major stream to capture another.
A. Beginning of downcutting by major streams and tributaries.
B. Stage just before capture period.
C. Diversion of one major stream by another as a consequence of capture.

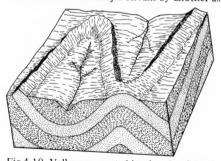

Fig.4-10. Valleys excavated in plunging folds in layered rocks (shale and sandstone).

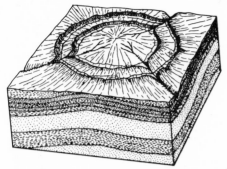

Fig.4-11. Valleys and hogbacks excavated in a dome formed by an intrusion (not shown) from below.

The complexity of drainage patterns and valley configurations in folded rocks increases as the complexity of folding increases. Examination of the cross-sections of several kinds of folds in Fig. 4-12 and consideration of the fact that fold troughs and crests may be either horizontal or plunging and straight or sinuous suggest the great number of possibilities that exist with regard to the distribution and kinds of rocks that may be encountered in downward erosion by streams in folded terranes.

FRACTURES IN ROCKS

Fractures are discontinuities resulting from failure of rocks under stress, either tensional or compressional. Faults are fractures along which there has been notable displacement of one side of the fracture relative to the other. A *joint* is a crack which transects a rock with little or slight displacement of adjacent sides. Faults generally are accompanied by joints, but joints are not necessarily associated with faults. Fracturing is of great concern in the evaluation of the engineering properties of rocks. Not only do fractures greatly reduce the strength of rocks so as to promote slope failure in valleys, but they also provide channelways for circulation of groundwater which expedite mechanical and chemical weathering and alteration by solutions of deep-seated origin.

A *fault* is a dislocation along a fracture as compared with a *fault zone*, which is characterized by movement along a number of parallel, subparallel, or intersecting slip surfaces. In many fault zones joints associated with faulting are numerous, and dislocation of the rocks is in part attributable to minor adjustments along joint surfaces.

The *strike* of a fault is the compass bearing of a horizontal line in the plane of the fault. The dip is the inclination of a fault from the horizontal and is measured in a plane at right angles to the strike. The *hade* is the inclination of the fault plane from the vertical. The wall above a fault is the *hanging wall*, as opposed to the

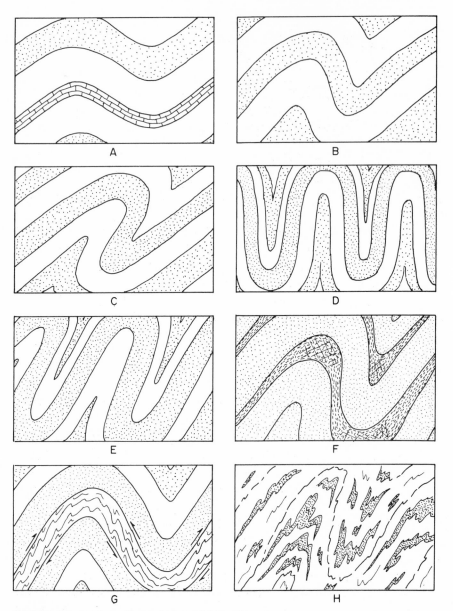

Fig.4-12. Cross-sections of several kinds of folds. No scale.
A. Symmetrical anticline and symmetrical syncline.
B. Asymmetrical anticline and asymmetrical syncline.
C. Overturned folds.
D. Isoclinal folds with vertical axial planes.
E. Isoclinal folds with inclined axial planes.
F. Incompetent material (shale) squeezed into crest of an anticline and trough of a syncline.
G. "Drag folds" in less competent layer between more competent layers. Arrows indicate movement of layers past each other during folding.
H. "Flow folds" in deformed, recrystallized metamorphic rock.

footwall, which is beneath the fault. The side of a fault that has moved relatively downward is the *downthrown* side. Conversely, the side that has moved relatively upward is the *upthrown side.* In *normal faults* the hade is toward the downthrown side, and in *reverse faults* the hade is toward the upthrown block. Reverse faults with high hade angles (and low dip angles) commonly are called *thrust faults.*

The relative displacement along faults of reference points on opposing sides of the fault surface permits recognition of *strike slip, dip slip,* and *oblique slip*

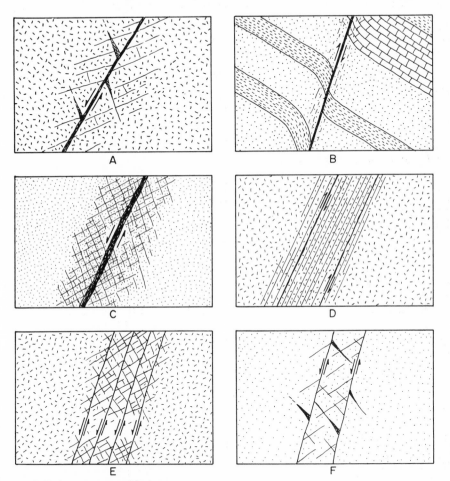

Fig.4-13. Some aspects of faults.
A. Tensional "gash jounts" and tight compressional "shear joints" have developed along a fault.
B. Faulting has caused "drag folding" in adjacent sedimentary rocks.
C. Fault is accompanied by extensive development of joints in wall rocks.
D. "Sheeted" fault zone. Movement along faults is accompanied by development of parallel joints.
E. Fault zone contains closely spaced joints.
F. "Gash joints" and "shear joints" have developed within and adjacent to fault zone.

faults. Faults intersecting pre-existing planar or linear structures are *discordant* or
cross-cutting faults. Accordant faults include *bedding-plane faults* and faults parallel
to foliation in metamorphic rocks.

Displacements along faults as measured by the relative dislocation of initially
adjacent reference points on opposing sides of a fault vary widely. Some faults are
the consequence of a single dislocation, but other faults have complex histories of
oscillatory, multidirectional displacements. The net measured displacement may
have a small magnitude, but if the fault has been the site of repeated dislocations,
the total displacement as measured by the total length of travel on one side or other
of the fault of a reference point may be considerable. Accordingly, the extent to
which rocks are crushed within a fault zone and the width of an associated zone of
jointing do not necessarily correlate with the displacement measured by matching
structures on opposing sides of the fault.

Faults commonly result in development of tensional or compressional joints
in their walls as indicated in Fig.4-13A, or cause adjacent rocks to be bent into
"drag folds", as in Fig. 4-13B. Fig. 4-13C—F illustrate other common aspects of
faults, fault zones, and associated joints.

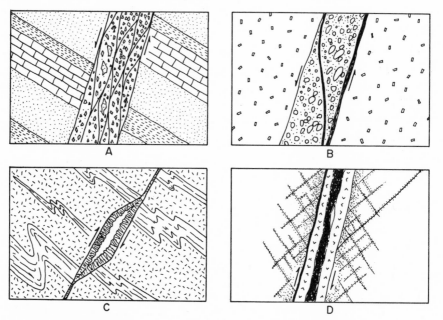

Fig.4-14. Materials in faults.
A. "Braided" slip surfaces intersect fault filling, consisting of angular "crush breccia".
B. Fault filled with gouge and "'crush conglomerate".
C. Open space in fault contains a partial filling of minerals deposited from solutions that moved
along the fault.
D. Fault contains a vein of gangue and ore minerals. Wall rocks are altered by the solutions that
deposited the fault filling.

Many faults and fault zones contain crushed materials and/or secondary minerals deposited from groundwater circulating along the faults (Fig.4-14). Fine-grinding of materials in faults produces *gouge,* which consists of a large proportion of grains of clay-size or slightly larger. *Crush breccia* (Fig. 4-4A) contains angular rock fragments usually embedded in gouge. Extensive dislocation along a fault may cause rounding of rock fragments between the walls of the fault to produce *crush conglomerate,* as in Fig.4-14B. Fig.4-14C and D illustrate faults in which introduced minerals partially to completely fill faults.

Joints are present in most rock bodies, large or small, and form as a consequence of a variety of natural processes. A distinction is made between *tension joints* and *shear joints,* depending on whether the joints develop by incipient or actual separation of adjacent walls under tension or whether the joints have formed by slight shear adjustments along planes of rupture. Actually, shear joints may be thought of as faults of almost zero to negligible displacement. Rupturing of rocks by tension jointing generally produces rough, irregular surfaces (Fig.4-15) that tend to prevent easy movement of adjacent joint blocks past each other, and slopes in bedrock containing such joints are more stable and stand with steeper inclinations than rocks containing smooth shear joints (Fig.4-15). Alteration along joints, whether by weathering or by hydrothermal solutions, commonly

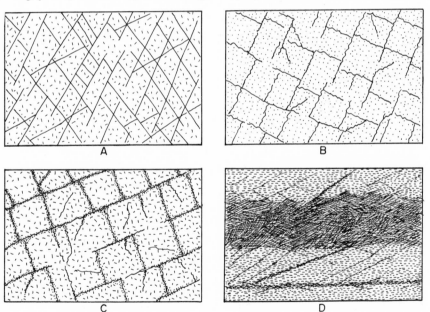

Fig.4-15. Shear joints and tension joints.
A. Smooth-surfaced, intersecting shear joints.
B. Rough-surfaced tension joints.
C. Tension joints that have localized alteration of wall rocks.
D. Closely-spaced, shear joints associated with displacements along bedding planes in shale.

introduces clay minerals which greatly reduce the frictional resistance between
adjacent joint blocks, whether the joint surfaces initially are smooth or rough.

Joints that are laterally and vertically extensive are sometimes called *master
joints,* and ordinarily are shear joints. Less persistent joints associated with master
joints are *subsidiary joints.* Parallel joints in a rock body comprise a *joint set* and
two or more intersecting joint sets constitute a *joint system.* The shapes of joint
blocks depend on the number and orientations of intersecting joint sets. When only
a single joint set is present, joint blocks tend to be *platy* or *slabby.* In rocks
containing two or more sets joint blocks have shapes that can be described as
prismatic, columnar, rhomboidal, cubic, blocky, or *irregular.*

Depending upon their origin, joints may be classified as *primary* or *secondary.*
Primary joints develop as a result of the processes active in a rock body at the time
of its formation or slightly later. Primary joints in sedimentary rocks include
contraction joints formed by dessication of clayey materials and joints of uncertain
origin in brittle layers as represented in Fig.4-16. The joints shown in Fig.4-16

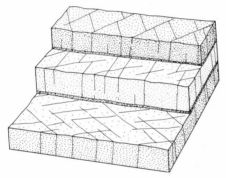

Fig.4-16. Joints in brittle, undeformed sedimentary layers. Note different directions of joints in
different layers.

commonly are present in limestones and well-cemented sandstones in horizontally-
layered sedimentary sequences. Typically they intersect the bedding at large to
right angles and display different directions in different layers. Such joints contribute notably to the permeability to circulating groundwaters of otherwise impermeable sedimentary layers.

Primary joints in igneous rocks commonly are contraction joints resulting
from volume decreases during cooling. In medium- to coarse-grained igneous rocks
containing platy, tabular, and prismatic minerals movement in magmas during
cooling and crystalization produces "flow layering" and "flow lineation", which is
seen in parallelism of inequidimensional mineral grains. Shrinkage during continued
cooling produces primary tension joints parallel and exactly or nearly perpendicular
to the flow layering as indicated in Fig.4-17. Because of the actual existence or the
tendency for the development of these joints many igneous bodies are characterized

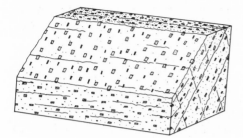

Fig.4-17. Joints in igneous rock showing "flow layering". Joints parallel to layering define the "rift" of the rock. The "grain" of the rock is parallel to joints intersecting the "rift" at right angles.

by a conspicuous joint system which yields small to large joint blocks. Another kind of primary jointing is seen in tabular or layered igneous bodies, either intrusive sills or thick lava flows. Contraction during cooling causes gradual extension of joints from the upper and lower surfaces toward the center of a tabular body to produce *columnar jointing* as indicated in Fig.4-18. Many of the columns are six-sided owing to intersection of shrinkage cracks at angles of 120°.

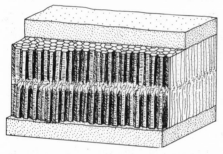

Fig.4-18. "Columnar jointing" in a sill of igneous rock intruded between sedimentary layers. Many of the columns are six-sided.

In field studies of jointed rocks a distinction should be made between *surficial joints* and joints which are present at depth and are not related to processes such as weathering and slope failure active at and near the earth's surface. Surficial joints generally are not present beneath the earth's surface at depths in excess of several tens to one or two hundred feet. Some examples of surficial joints are given in Fig.4-19. In Fig.4-19A surface solutions have penetrated a rock along already existing joints and have caused the development of concentric weathering joints by alteration of the rock materials in joint blocks. Weathering alteration of this kind is called *spheroidal weathering*. In Fig.4-19B expansion and contraction of an exposed rock surface in response to short- and long-term temperature changes has produced a set of joints parallel to the exposed surface. Separation of slabs bounded by joints is expedited by weathering along the joints.

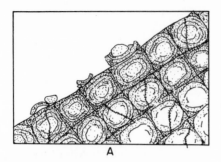 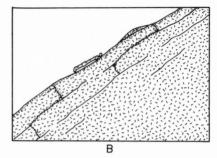

Fig.4-19. Joints in surface exposures of rocks.
A. "Spheroidal weathering" develops joints with curved surfaces by alteration localized by previously existing joints.
B. Joints have developed parallel to exposed surface by expansion and contraction. Weathering along joints is indicated by stippling.

FRACTURES ASSOCIATED WITH FOLDS

Folding in rocks is a very complex process involving dislocations of a variety of kinds on both microscopic and megascopic scales. The particular behavior of a rock during folding is closely dependent on its properties, particularly its strength. On a megascopic scale rocks in folds are deformed by plastic flow, elastic dislocations, recrystallization, and development of tension and shear fractures. Commonly a close association exists between fold anisotropism and fracture anisotropism in rocks, and both kinds of anisotropism are the consequences of the same general kinds of stresses acting to deform the rocks.

Faults eventually terminate vertically and horizontally in zones of decreasing fracture dislocation of rocks, at the earth's surface, or in folds. Crystalline basement rocks below folded sedimentary rocks generally contain faults or intricate joint systems that were formed by adjustments that were translated into folds in the overlying sediments, and many fold displacements laterally or vertically become fault displacements. A typical fold—fault relationship is shown in Fig.4-20, where a sharply folded sedimentary sequence passes downward into a fault.

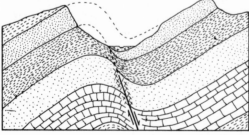

Fig.4-20. Folded rocks are displaced by a fault at depth. Folds in contact with fault are "drag folds".

Brittle, strong rocks, such as limestone and well-cemented sandstone rupture more easily during folding than softer rocks such as shale and mudstone. During folding layers move past each other in the limbs of folds and in layers of strong rocks stretching at the tops of anticlines and at the bottoms of synclines to produce tension cracks and compression in the inside portions of folds causes generation of shear joints (Fig.4-21). Movement of brittle layers past each other, as in Fig.4-21,

Fig. 4-21. Diagrammatic representation of relationship between folds and associated fractures. Arrows show relative movement of layers during folding along bedding-plane faults. Tight shear fractures and open tension fractures have formed where indicated. Shale layers between brittle competent layers have been squeezed into crests of anticlines and troughs of synclines.

causes development of both tension and shear joints. Softer materials such as shale in Fig.4-21 tend to be displaced by plastic movement and shearing into anticlinal crests and synclinal troughs.

Folding during metamorphism of layered rocks is a consequence of either "shear folding" or "flow folding" or a combination of the two. In shear folding folds develop by slight displacements along closely spaced parallel surfaces as in Fig.4-22A. Commonly there is some recrystallization of platy or micaceous

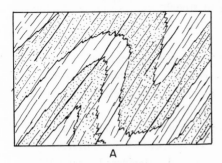
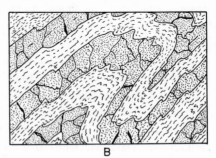

A   B

Fig.4-22. Joints in folded metamorphic rocks.
A. Joints develop along "slaty cleavage" in metamorphic rock containing "shear folds".
B. Irregular joints in brittle layers between mica-rich layers in irregularly folded metamorphic rock.

minerals in the direction of shearing, and the combination of shear planes and the directional orientation of recrystallized minerals promote easy splitting and development of joints. The tendency to split in the direction of the planar elements is described as *slaty cleavage*, and is exemplified by the easy parting along parallel planes in slate.

Recrystallization and deformation in layered metamorphic rocks in "flow folding" produces intricate, complex folds, as in Fig.4-22B. In Fig.4-22B layers rich in mica alternate with brittle layers containing quartz and/or feldspar. Dislocation of the rocks has produced highly irregular joints in the brittle layers but not in the mica-rich layers.

VALLEYS IN FRACTURED ROCKS

Faults and strong joint sets exert a major influence in determining the courses of streams as they excavate their valleys. In addition, joints, whether associated with faults or not, contribute substantially to the failure of slopes in valleys. Recognition of the fact that a valley has been eroded in a fault or in strongly jointed rocks is of paramount concern in the investigation of dam and reservoir sites because of the generally low strength of the fractured rocks and because of the distinct possibility that the fractured rocks may be highly permeable to the movement of subsurface water.

Valleys that have straight courses for substantial intervals commonly follow linear zones of weak bedrock, and such valleys should be investigated very carefully to determine whether the weak rock is localized by a fracture zone or bedrock is inherently weak because of its original physical characteristics. Linear valleys develop in soft layers in tilted sedimentary rocks, but geological investigations usually reveal the origin of such valleys. In the search for fractured zones in bedrock special attention should be given to straight valleys that are discordant with respect to bedrock structure or, in crystalline rocks, are not in harmony with a regional drainage pattern.

In areas of valley glaciation originally sinuous stream-cut valleys tend to be straightened by removal of valley spurs by glacial erosion, but, unless it can be demonstrated that the valley was notably sinuous prior to glaciation, the presence of a fractured zone beneath the valley floor should be suspected.

Fig.4-23 and 4-24 illustrate valleys in fault zones in crystalline rocks such as granite and metamorphic schist and gneiss. These rocks in the unfractured condition generally are highly resistant to erosion, and differences in erodability are not great. In Fig.4-23 a symmetrical straight valley has followed a steeply dipping zone of fracturing. Tributaries to the trunk stream in the fault-controlled valley have sinuous courses and have excavated valleys with no structural controls in

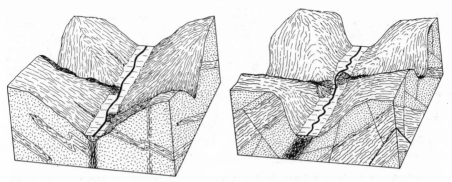

Fig.4-23. Symmetrical valleys excavated in crystalline igneous and metamorphic rocks. A straight major valley containing a trunk stream has been eroded in a vertical fracture zone. Valleys of tributary streams are sinuous and exhibit no structural control in bedrock.

Fig.4-24. Straight asymmetrical glaciated valley formed by glaciation of an asymmetrical stream-cut valley in an inclined fracture zone. Glaciated tributary valleys are symmetrical and preserve original symmetry.

bedrock. In Fig.4-24 it is assumed that a major stream eroded an asymmetrical valley along an inclined fracture zone and that the stream valley was subsequently modified by a valley glacier. Although the main valley retains the original asymmetry of the stream-cut valley, glaciated tributary valleys not localized along fracture zones have developed symmetrical cross-sections.

Fig.4-25 and 4-26 show valleys excavated in fault zones intersecting alter-

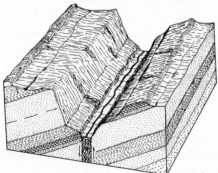

Fig.4-28. Steep slopes in a canyon in moderately fractured crystalline schists and gneisses. Blue Mesa Dam and powerplant on the Gunnison River, Colorado. (Photo courtesy of U.S. Bureau of Reclamation.)

nating hard and soft sedimentary rocks. In Fig.4-25 the fault zone is parallel to the average strike of the sedimentary layers, that is, it is a *strike fault*, but it is not parallel to the bedding. A straight valley excavated in the fault zone is parallel to hogbacks and strike valleys in unfaulted rocks on both sides of the valley, but a mismatch of layers across the valley indicates that a fault underlies the valley and not a soft sedimentary layer. In Fig.4-26 the fault intersects bedding obliquely and

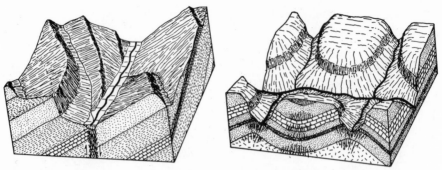

Fig.4-26. A steeply-dipping fault zone obliquely intersects tilted sedimentary rocks and produces a straight valley which is discordant with the layers in the sedimentary rocks.

Fig.4-27. Streams tributary to a discordant trunk stream have cut valleys in fractured rocks in crests of parallel, symmetrical anticlines. Valleys have been eroded below an original surface of low relief.

produces a valley which is discordant with respect to the layering in the sedimentary rocks. The presence of a fault in bedrock is further indicated by a mismatch of beds on opposing sides of the valley.

Fig.4-25 and 4-26 represent only two possibilities of a great many. Stream excavation in fault zones intersecting sedimentary rocks can produce simple or complex patterns of erosion, both in plan and cross-section, depending on whether bedrock structure is simple or complex. Maximum complexity is associated with folded rocks intersected by numerous faults of different directions and attitudes and with differing amounts of fault displacement and associated jointing.

Joints, unless they exist in single sets of great lateral persistence, rarely determine the courses followed by streams in major valleys. A notable exception is seen in the diagrammatic representation in Fig.4-27, which illustrates the controls in valley dissection by fractures, predominantly joints, in the crests of parallel anticlines. The major stream and the tributary streams have cut valleys below an original surface of low relief. The course of the major stream was initially determined by the slope of the original surface, over which it flowed, and the stream has maintained its course in spite of the structures encountered in bedrock. The tributary streams, however, developed valleys by headward and downward erosion in the fractured rocks in the crests of the anticlines.

SLOPE DEVELOPMENT IN ANISOTROPIC BEDROCK

A typical valley in anisotropic rocks develops by a combination of downward erosion and slope failure, which, together, produce a profile that reveals subtle to gross differences in bedrock. Weathering and groundwater movement tend to accelerate slope failure, but development of steep, even vertical slopes, alternating with flatter slopes, ultimately depends on the physical and mineralogical character-istics of individual rock bodies.

Elements of primary anisotropism that contribute to the behavior of bed-rocks in slopes are planes of physical discontinuity or separation, such as bedding planes in sedimentary cocks, planes of mineralogical discontinuity as in layered metamorphic rocks, particle size, distribution, and orientation, and manner of articulation of mineral grains and/or rock fragments. Additional factors that must be taken into account are differences in permeability, porosity, and mineral composition in various rock units in anisotropic assemblages. The influence of secondary anisotropism in contributing to slope development in bedrock depends on the nature of the anisotropism, or directional anisotropism associated with rock alteration. Fracturing exerts a particularly spectacular influence on slope develop-ment in producing surfaces of physical discontinuity and in providing avenues for circulation of groundwater which alters the rocks chemically, develops pore pres-sures, or expands by freezing.

The effectiveness of fracturing, either faulting or jointing, in promoting rock disintegration or decomposition is a function of spacing and attitude. Closely spaced shear planes or closely spaced joints in one or several sets reduce bedrock to a relatively incoherent material that cannot stand in steep slopes, whereas widely-spaced fractures tend to control detachment of large blocks or slabs from bedrock exposures in a manner that may lead to development of precipitous slopes.

Joints develop as a result of faulting, by tensional and compressional forces associated with folding, and by processes associated with compaction and/or dessication of sedimentary layers of appropriate composition and fabric. In the last category are vertical joints that are seen in undisturbed, horizontal layers of brittle, high-strength sandstone and limestone and microjoints that intersect the bedding of fissile shaly rocks and presumably formed as a consequence of dessication after burial.

Steepest slopes (up to vertical) develop in relatively unfractured rocks of moderate to high strength (Fig. 4-28). Slopes of least inclination form on soft, poorly-cemented bedrocks containing abundant clay minerals. Slopes on bedrock, either exposed at the surface or buried beneath a mantle of unconsolidated debris, as related to bedrock characteristics are summarized in Table 4-1. Implicit in the summary is the assumption that weathering, either chemical or mechanical, is more effective in slope reduction in closely fractured rocks than in massive or only slightly fractured rocks.

Considerably less widespread than weathering alteration of rock is alteration by warm to hot solutions of deep-seated origin. Hot solutions, called *hydrothermal solutions,* have caused extensive changes in bedrocks in volcanic areas and in many areas of economic mineralization. A common result of hydrothermal alteration is development of clay-mineral aggregates and reduction of the strengths and ability to stand in steep slopes of initially competent rocks.

Fig.4-28. Steep slopes in a canyon in moderately fractured crystalline schists and gneisses. Blue Mesa Dam and powerplant on the Gunnison River, Colorado. (Photo courtesy of U.S. Bureau of Reclamation.)

TABLE 4-1

Valley slopes in bedrock

I.   Slopes in unfractured to moderately fractured rocks. (Steep slopes develop more con-
      sistently in rocks in which fractures are nearly or exactly vertical and/or essentially
      horizontal.)
   A. Steep slopes with inclinations from about 45° to vertical.
      1. Igneous rocks in intrusive bodies and lava flows.
      2. Metamorphic rocks including slate, schist, gneiss, migmatite, hornfels, marble, and
         quartzite.
      3. Sedimentary rocks.
         a. Pyroclastic rocks with firmly welded, interlocking particles as in *welded tuffs*
            and some *agglomerates.*
         b. Clastic silicate rocks with abundant interstitial cement of silica, iron oxides, or
            carbonates, including some conglomerates, sandstones, siltstones, shales, mud-
            stones, and claystones.
         c. Limestone and dolomite, especially if partly to completely recrystallized.
   B. Intermediate slopes with inclinations from about 25° to about 45°.
      1. Sedimentary rocks.
         a. Partly indurated pyroclastic rocks.
         b. Sedimentary rocks with little or no interstitial cement. Conglomerates, sand-
            stones, and siltstones with clay mineral matrix. Compact shale, mudstone, and
            claystone.
         c. Chalk or shell limestones and dolomites.
   C. Flat slopes with inclinations of about 5° to about 25°.
      1. Sedimentary rocks.
         a. Pyroclastic rocks including tuff, volcanic-ash deposits, and agglomerate and
            consisting of loose, relatively uncohesive particles. Pyroclastic rocks that have
            been altered, at least in part, to swelling clay minerals (montmorillonite) as in
            bentonite are particularly unstable in slopes when brought into contact with
            water and assume very flat slopes.
         b. Soft shales, mudstones, and claystones rich in clay minerals. If swelling clays
            (montmorillonite) are present, very flat slopes are generated.

II.   Slopes in closely fractured rocks.
   A. Steep slopes with inclinations of about 45° to vertical. Nonexistent unless fractures
      contain introduced cementing materials which increase competency.
   B. Intermediate slopes with inclinations of about 25° to about 45°. Steeper slopes in
      rocks containing vertical and/or horizontal fractures.
      1. Igneous rocks in intrusive bodies and lava flows.
      2. Metamorphic rocks including slate, schist, gneiss, migmatite, hornfels, marble, and
         quartzite. Biotite-rich rocks tend to weather easily and contribute to development
         of slopes of low inclination.
      3. Sedimentary rocks.
         a. Pyroclastic rocks with firmly welded, interlocking particles.
         b. Clastic silicate rocks with strong cement binding grains together.
         c. Limestones and dolomites, especially if partly to completely recrystallized.
   C. Flat slopes with inclinations of about 5° to about 25°.
      1. Igneous rocks, especially if thoroughly altered by chemical weathering.
      2. Metamorphic rocks, especially if thoroughly weathered.
      3. Sedimentary rocks.
         a. Pyroclastic rocks, especially when they contain abundant clay minerals.
         b. Clastic sedimentary rocks with clay-mineral matrix, including conglomerate,
            sandstone, and siltstone. Soft shale, mudstone, and claystone.
         c. Limestones and dolomites such as chalk and shell limestone characterized by
            weak bonding between particles.

SLOPE FAILURE DEPENDENT ON ROCK ANISOTROPISM

Primary anisotropism, such as foliation and bedding, and secondary anisotropism resulting from folding and fracturing under many circumstances contribute to slope instability in valleys. Fracturing, particularly, promotes weathering disintegration and decomposition of bedrocks, and slopes that may not have failed during

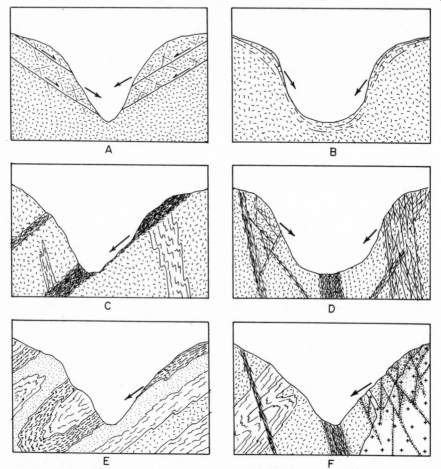

Fig.4-29. Unstable slope configurations in igneous and metamorphic rocks. Arrows indicate location and direction of movement during failure.
A. Gravity-slip faults dip toward valley and induce slope instability.
B. Joints subparallel to rock surface in a glaciated valley have formed by long-term temperature changes, and, probably, elastic rebound.
C. Materials in a flat fault zone are perched precariously on one side of the valley.
D. Fractured rocks in valley walls have greatly reduced rock competency in a glaciated valley.
E. A layer of mica schist in folded metamorphic terrane presents a potential hazard.
F. Thoroughly fractured and altered rocks associated with intrusive igneous stock produce rocks of low competency capable of slope failure.

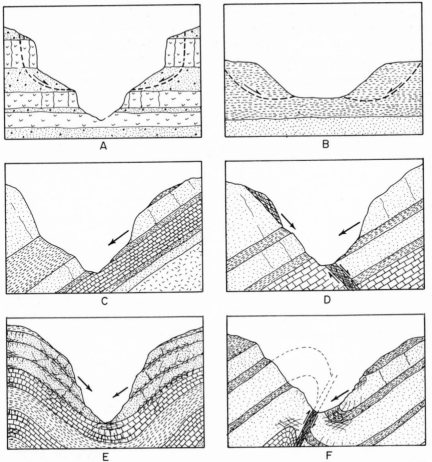

Fig.4-30. Unstable slope configurations in layered rocks. Arrows indicate locations and directions of movement of bedrock materials during slope failure. Potential planes of failure are indicated by dashed lines.
A. Lava flows alternate with beds of volcanic ash (stippled).
B. Slopes in horizontal shale layers, especially if they are weathered, may fail as indicated.
C. A strike valley in inclined sediments contains a layer of sandstone above a layer of shale creating a particularly unstable slope configuration.
D. Faulted, tilted sedimentary rocks have a potential for slope failure.
E. Fractures resulting from slippage of layers past each other have reduced competency of brittle rocks.
F. Folded and faulted rocks present potential for slope collapse.

an earlier stage of valley excavation become increasingly unstable as weathering proceeds. Thus, valley excavation and slope reduction are usually intermittent processes. Recognition of precariously unstable slopes that may fail as a consequence of dam construction and reservoir filling or that may fail spontaneously during the lifetime of a dam and/or the reservoir it impounds is imperative. Early recognition of potentially hazardous conditions in bedrock beneath valley slopes may eliminate further consideration of a possible site for a dam and reservoir.

Fig.4-29 and 4-30 illustrate several potentially hazardous conditions in bedrock in the walls of valleys. The reader will understand that only a few of a great many possible kinds of undesirable bedrock conditions are shown in the illustrations. Fig.4-29 shows several kinds of unstable slope configurations in crystalline igneous and metamorphic rocks. Fig.4-30 indicates several examples of potentially dangerous conditions in layered rocks, including sedimentary rocks and volcanic rocks.

Chapter 5

GROUNDWATER HYDROLOGY OF UNIFORMLY PERMEABLE MEDIA

INTRODUCTION

This chapter is concerned with the flow of water (seepage) beneath dams and around abutments. In addition, consideration is given to the effects on stability of penetration by water of natural materials within and around the periphery of a reservoir impounded by a dam. In the development of basic concepts a simplifying assumption will be made that the materials under consideration have essentially uniform permeabilities vertically and horizontally. Discussion of anisotropic conditions that control groundwater movement are deferred to a later chapter (Chapter 6). Also, the special conditions associated with penetration by water of soluble rocks such as limestone, dolomite, and gypsum are given separate consideration in a subsequent chapter (Chapter 7).

Ideally a dam should be constructed of completely impermeable materials and should be located on completely watertight rock or soil in the foundation and abutment areas. As a practical matter, materials commonly used in dam construction, such as earth and concrete, are not completely impermeable, and foundation rocks and soils show wide ranges in primary and/or secondary permeability. Control of the permeability of materials emplaced during construction is much simpler than reduction of undesirable permeability to a safe level beneath or in the vicinity of a dam. Accordingly, a thorough investigation of bedrock geology and groundwater hydrology of the site of a proposed dam is essential before design and construction, so that adequate precautionary measures may be taken during construction to provide an acceptable level of safety.

Within the dam foundation and abutments seepage results in water loss by leakage, uplift forces at the bottom of the dam, possible creation of instability in foundation and abutment rocks or soils, and, in some instances, development of destructive high-velocity concentrated flows by piping and cavitation. A limited amount of slow seepage through and below a dam is permissible, but every effort must be made to control and/or prevent concentrated seepage flows that will threaten the integrity of the foundation or the dam above it.

When a body of water is impounded in a reservoir by a dam, marked changes in the hydrologic environment on the floor and sides of the body are inevitable. Not only is seepage from the reservoir a distinct possibility, but water-soaking of permeable unconsolidated materials within and around the periphery of the reservoir can create unstable slopes that might collapse with disastrous or at least highly undesirable consequences.

Accordingly, routine study of a site for a proposed dam and reservoir should include careful investigation of both the dam foundation and the storage basin behind it.

SOME DEFINITIONS

Groundwater, as the name implies, is water below the earth's surface, either below standing or moving bodies of water or below the ground–atmosphere contact. Flow of water underground requires *permeability,* and permeable rocks and soils are designated as *aquifers.* Bodies of soil or rock of very low permeability, through which water moves very slowly, or not at all, are called *aquicludes.* A rock in which a primary permeability is essentially absent is described as an *aquifuge.*

Water below the surface is found in a *zone of aeration,* where interstices are filled with both air and water, and a *zone of saturation* in which only water is present in pore spaces. The top of the saturated zone is the *water table* (Fig.5-1)

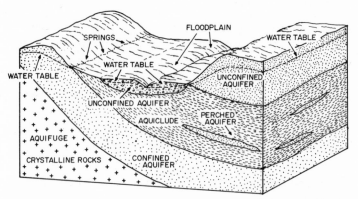

Fig.5-1. Block diagram illustrating several commonly used geohydrologic terms.

and is a surface where the pressure in the water is equal to atmospheric pressure. When interstices are of appropriate size to enable capillary forces to operate, a *capillary fringe* forms above the water table.

Groundwater flow through a *confined aquifer* bounded by impermeable surfaces as in a sandstone between shale layers is *confined flow.* Flow of water through an *unconfined aquifer* below the groundwater table, which commonly is designated as a *free surface,* is *unconfined flow.* Confined and unconfined aquifers are shown diagrammatically in Fig.5-1.

PROPERTIES OF WATER

   The physical state of water determines its behavior above and below the earth's surface. The water molecule has a radius of about 1.42Å and, because of the asymmetrical coordination of hydrogen and oxygen in the molecule, each water molecule acts as a dipole with positive and negative-electrical charges on opposing sides. Mutual attraction of the dipoles accounts for the high boiling point (100°C), a high dielectric constant (80 at 20°C), and a high surface tension (72.7 dynes/cm against air), as compared with most other liquids.
   Because of the opposed electrical charges, the water molecule is strongly attracted to the surfaces of solids, especially solids with unsatisfied surface charges. The force of attraction to a solid surface coupled with a high surface tension produces capillary forces of large magnitude.
   Of concern in the design of dams and in the study of groundwater hydrology are the density and viscosity of water as functions of temperature. Fig.5-2 shows

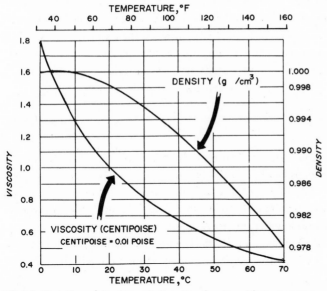

Fig.5-2. Variation of density and viscosity of water with temperature.

how density and viscosity vary with temperature. Within the ranges of pressures of present concern water can be considered as virtually incompressible, and density changes as a function of temperature are much more significant than density changes attending increased confining pressure.
   Fluids, including water, flow by means of shear dislocations of very small volumes relative to each other. Opposing these dislocations is fluid friction, which depends on the intermolecular attraction between the molecules. Resistance to

fluid flow is called *viscosity* and is defined as the ratio of the shearing stress to the rate of shear. Water is essentially a Newtonian liquid in which viscosity is independent of the rate of shear over a wide range of shear rates.

If the shearing stress is measured in dynes/cm^2 and the rate of shear in reciprocal seconds (sec^{-1}), the viscosity is given in dyne sec cm^{-2}, or, equivalently, as g sec^{-1} cm^{-1}. The unit of viscosity is the *poise*. The unit of *kinematic viscosity*, which takes into account the density of the liquid, is the *stokes* and is equivalent to the viscosity in poises divided by the density of the liquid in g/cm^3. For water the ordinary viscosity is essentially equal to the kinematic viscosity (density of water = unity). Viscosity of water changes with temperature as shown in Fig.5-2. The unit in Fig.5-2 is a *centipoise*, which is 0.01 poise.

Viscosity may be expressed in various ways, depending on the units that are used in the equation for viscosity. In engineering practice a commonly used unit is 1b ft^{-1} sec^{-1} which is equivalent to 14.882 poise (g cm^{-1} sec^{-1}).

At small velocities flow of water commonly appears to be in layers moving past each other, as in a pipe or an open conduit. This type of flow is described as *laminar*. With increasing velocity the pattern of flow becomes irregular and assumes the characteristics of *turbulent flow*. A useful criterion for estimation of whether flow through a permeable medium is laminar or turbulent is the *Reynolds number*, which has found wide application in the study of fluids in motion.

The Reynolds number R for permeable soils is expressed as:

$$R = VD\rho/\mu \tag{5-1}$$

where V is the discharge velocity (cm/sec), D is the average diameter of soil particles (cm), ρ is the density of the fluid (g/cm^3), and μ is the viscosity (g sec^{-1} cm^{-2}). Experimental study of the change in laminar flow to turbulent flow in soils yields critical Reynolds numbers of from less than 1 to 7. In most soils the velocity of movement of water is so low that the Reynolds number is 1.0 or less.

If we assume that the maximum (critical) Reynolds number for laminar flow is 1.0 or less and that the viscosity and density of water are both unity we arrive at:

$$R_{subcritical} = VD \lesssim 1.0 \tag{5-2}$$

As an example suppose that water is percolating with a velocity of 0.5 cm/sec through a clean sand consisting of particles with diameters near 0.2 cm. Then:

$$R = 0.5 \times 0.2 = 0.1 \tag{5-3}$$

and the Reynolds number clearly is consistent with laminar flow.

The velocity of flow, V, in the above equations refers to flow on a macroscopic scale and is the average velocity through a given cross-section. It is *not* the velocity of flow through the intricately interconnecting voids of complex shape in a permeable medium that are observed on a microscopic scale.

POROSITY AND PERMEABILITY

Porosity is the percentage of a given volume of a substance occupied by voids. Free flow of water through a substance is possible only when the voids are interconnecting and the substance has *effective porosity* and can be described as *permeable.* Water flows easily through clean sands and gravels, but with difficulty through very fine-grained clay and shale, although the porosity of the shale may greatly exceed that of the clean sand.

Permeability may be a consequence of the original manner of formation of a material and is then described as *original permeability.* Examples of original permeability are found in sedimentary rocks which are built up layer-upon-layer by deposition of mineral or rock fragments and/or chemical precipitates. *Secondary permeability* results from the development of fractures in brittle materials and from dissolution by solutions penetrating the materials. Examples of secondary permeability are observed in jointed rocks of all kinds and in limestones and dolomites that have been rendered porous or even cavernous by solutional activity of moving groundwaters.

LAMINAR FLOW OF WATER THROUGH PERMEABLE MEDIA

The quantity, Q of water that passes through a unit area of cross-section in unit time is the *discharge velocity,* sometimes also identified as the *discharge quantity,* the *quantity of seepage,* or as the *discharge.* The following relationship relates Q to the ratio of pores, p, to solids, s, in the area of cross-section, A, and to V, the so-called *seepage velocity.*

$$Q = pAV/s \qquad\qquad\qquad (5\text{-}4)$$

The driving force that causes water at the earth's surface to penetrate permeable materials below the surface ultimately is gravity. The pressure exerted by water at any point or on any plane is the *hydraulic head.* In standing bodies of water the pressure as a function of depth y is the *hydrostatic head.* The unit of hydrostatic (pressure) head is the pressure exerted on a unit area by the mass of a unit volume of water. A unit of hydrostatic head commonly is taken as the weight of a cubic foot of water (62.37 lbs) on an area of a square foot. This unit reduces to

0.433 lbs/inch^2 (at 60°F). In the c.g.s. system the unit of head is the weight of a cubic centimeter or a cubic meter. The calculation of *elevation head* in a standing body of water is simply accomplished by multiplying the unit of pressure head by a depth in feet, centimeters, or meters.

Water in motion has momentum and is characterized by a velocity head, which is expressed as $V^2/2g$, where V is the velocity and g is the acceleration due to gravity.

A general expression for head, H, in a fluid moving through a permeable substance is:

$$H = y + p/\gamma + V^2/2g \tag{5-5}$$

in which y is the elevation of a given point relative to an arbitrarily assumed horizontal reference plane, p is the pressure at the point, γ is the unit weight of the water, V is the velocity, and g is the gravitational constant, which can be taken as 980 cm/sec^2 or as 32 ft/sec^2. As a practical matter, seepage velocities generally are so low that the term $V^2/2g$ can be neglected, and we have:

$$H = y + (p/\gamma) \tag{5-6}$$

Laminar flow also can be characterized as *streamline flow,* especially when considered in a vertical, longitudinal section. Long ago Darcy (1856) established a relationship valid for laminar flow now known as *Darcy's law,* which is:

$$Q = -kA(\mathrm{d}h/\mathrm{d}s) \tag{5-7}$$

and, in differential form, states that the discharge velocity across a cross-section of area A is equal to a constant k multiplied by a term involving the head, h, and a measure of length, s, along a streamline. A constant, k, for each permeable medium is the *coefficient of permeability*; $\mathrm{d}h/\mathrm{d}s$ is the *hydraulic* or *piezometric gradient* and characterizes the head at a point on a streamline in steady laminar flow. For turbulent flow a general expression is:

$$Q = kA\,(\mathrm{d}h/\mathrm{d}s)^m \tag{5-8}$$

in which m is a number such that $0.5 < m < 1.0$.

The coefficient of permeability also has been designated by various investigators as the *seepage coefficient, effective permeability,* and *hydraulic conductivity.* Hydraulic conductivity also sometimes is considered to be a velocity along a streamline as measured in inches, feet, centimeters, or meters per unit of time.

Three commonly used coefficients relating to the permeable characteristics of a medium are:

(1) cm/sec or meters/sec (c.g.s. system). Volume is measured in cubic centimeters or cubic liters and cross-sectional area in square centimeters or square meters.

(2) *Meinzer unit (K)*. The flow of water in gallons per day through a cross-sectional area of one square foot under a hydraulic gradient of unity (100%) at 60°F. Expressed as gallons per day per ft^2 (gpd/ft^2). One meinzer is equivalent to 0.055 darcys for water at 60°F.

(3) *Darcy*. The equation for the darcy unit is:

$$\text{One darcy} = \frac{(\text{one centipoise}) \, (\text{one cm}^3/\text{sec})/(\text{one cm}^2)}{\text{one atmosphere/cm}}$$

One darcy is equivalent to 18.2 meinzer units for water at 60°F. One millidarcy = 0.001 darcy.

For convenient use in calculations Darcy's law may be written as:

$$Q = KHA/L \tag{5-9}$$

in which Q is the quantity of water flow across a cross-section of area A, K is a coefficient of permeability, H is the head loss per unit of length in the direction of flow, and L is the total length in the flow direction. When Q is expressed as gallons per day, and A as one square foot, K becomes Meinzer's unit.

Table 5-1 shows orders of magnitude of primary permeabilities of several natural materials expressed in meinzers and darcys. The permeabilities for the rocks assume that no fractures or channelways are present which might produce secondary permeabilities of small to large magnitude.

TABLE 5-1

Orders of magnitude of original permeabilities of some natural materials

	Meinzers	Darcys
Unconsolidated sediments		
Gravel	$10^4 - 10^6$	$10^3 - 10^5$
Clean sand	$10 - 10^4$	$1.0 - 10^3$
Clayey sand	$10^{-2} - 10$	$10^{-3} - 1.0$
Clay	$10^{-4} - 10^{-2}$	$10^{-5} - 10^{-3}$
Sedimentary rocks		
Sandstone	$10^{-6} - 10^{-2}$	$10^{-7} - 10^{-3}$
Shale	$10^{-6} - 10^{-4}$	$10^{-7} - 10^{-5}$
Limestone	$10^{-5} - 10^{-2}$	$10^{-6} - 10^{-3}$
Crystalline, igneous, and metamorphic rocks	$10^{-6} - 10^{-4}$	$10^{-7} - 10^{-5}$

HYDRODYNAMIC FLOW NETS BENEATH DAMS

Flow of water through permeable materials is directional and is in response to head (pressure) differentials. Flow can be graphically portrayed by hydrodynamic flow nets, which are usually drawn in vertical section parallel to the general direction of flow. A *flow net* is a graph containing two families of intersecting curves. One family consists of *flowlines* or *streamlines,* which are the loci of the paths of flow of individual particles of water. The other family consists of *equipotential lines,* which pass through points of equal head (or pressures). All intersections between the streamlines and equipotential lines are at right angles.

Accurate construction of flow nets by analytical methods requires satisfaction of Laplace's equations and is possible only when the geometry of the boundary conditions is simple and symmetrical, and the hydrodynamic parameters are known. In general, because of the complexity of analytical solutions under most natural conditions of groundwater flow a variety of approximate methods and model studies are substituted for analytical studies (Davis and DeWiest, 1966; Harr, 1962; Pietraru, 1968; Chugaev, 1971).

In the absence of solutions based on mathematical analyses, streamlines and equipotential lines may be sketched in a section beneath a dam, provided that the boundary conditions are at least approximately known. Nets constructed in this manner, although quantitatively imprecise, serve a useful purpose in identifying orders of magnitude of the influence of various factors on the flow of groundwater. Useful in making sketches is the fact that any two seepage zones that are geometrically similar will have similar flow nets, so that reference may be made in construction of a net to flow nets previously determined by analytical methods, approximate methods, or model studies.

The configuration of a flow net does not depend on the coefficient of permeability or on the initial head. However, permeability and head are important considerations in estimating the amount (velocity) of flow beneath a dam.

The locus of a streamline is defined by a usually unknown function $\psi(x,y) =$ a constant. Each streamline is identified by a separate constant, different from the constants for the other streamlines. Flow between any two arbitrarily selected streamlines, along a so-called *flow channel,* is also constant. Equipotential lines (curves of constant head or pressure) are generally unknown functions $\phi(x,y) = a$ succession of constants. The potential or head drop is everywhere the same in the flow area between adjacent equipotential lines.

Analytical solutions for $\psi(x,y)$ and $\phi(x,y)$ for the simple configuration of dam and subsurface in Fig.5-3 have been reviewed by Harr (1962) and others. These solutions provide the basis for quantitative assessment of the influence of various hydrodynamic parameters on subsurface flow of water through a homogeneous layer of infinite depth beneath a dam with its base at ground level.

The function $\phi\,(x,y)$ = a constant, by substitution of a succession of different constants generates a family of half ellipses confocal at foci at A and B. The equipotential curves ϕn are symmetrically disposed hyperbolas, each with its own constant. The hyperbolas in Fig. 5-3 are drawn so that $\Delta\,\phi$, the potential drop between adjacent curves is a constant equal to one-tenth of the total drop between $\phi = -kh$ and $\phi = 0$.

For an initial head, H = 50 ft and $2b$ = 40 ft (Fig.5-3), the pressure (head)

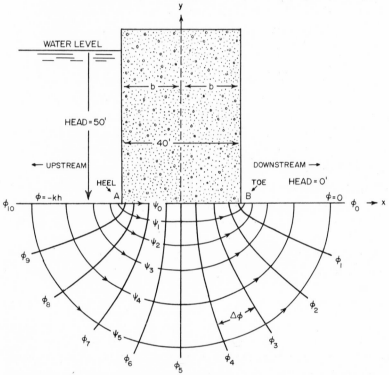

Fig.5-3. Symmetrical hydrodynamic flow net beneath a dam with its base at ground level. The material below the dam is homogeneous and of infinite depth. Streamlines are identified as ψ_0, ψ_1, etc., and equipotential lines as ϕ_0, ϕ_1, etc.

distribution at the base of the structure is shown graphically in Fig.5-4. Of concern in the design of dams is the so-called *uplift pressure*, which is the pressure exerted on the base of the dam by the pore water in a permeable medium beneath the dam. Because this pressure is a maximum at the heel of the dam (A) and diminishes to zero toward the toe (B) it tends to rotate the dam upward about an axis at the toe.

In engineering practice this tendency is analyzed by moments of force or torque. In a general expression the torque L is related to the force F acting to produce rotation about a center at a distance d from the point (or line) of application of the force by the relationship:

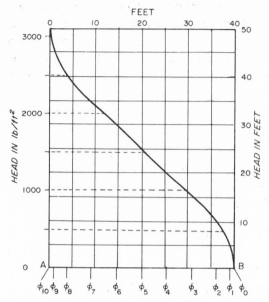

Fig.5-4. Distribution of pressure (head) at the base of the structure in Fig.5-3.

$$L = Fd \tag{5-10}$$

In Fig.5-3 the moment of the uplift forces M_A relative to the heel of the dam (A) as obtained by integration between the limits $-b$ and $+b$ is:

$$M_A = PX = 3bP/4 \tag{5-11}$$

in which P is the pressure per unit length acting on the base of the structure and X is the moment arm of the pressure force measured in the x direction from point A.

The velocity of flow at the $\Phi = 0$ boundary (downstream) is given by:

$$V = kh/\pi \, (x^2 - b^2)^{\frac{1}{2}} \tag{5-12}$$

in which h is the initial head equal to H and k is the coefficient of permeability. This relationship was used to construct the graph in Fig.5-5. The coefficient k is expressed in Meinzer units (gpd/ft^2). Three values are arbitrarily assigned to k:10, 10^{-3} and 10^{-5}, corresponding to the permeabilities of a clean sand, a sandstone, and a shale, respectively, and provide a comparison of volume flows through highly permeable, moderately permeable, and almost impervious materials.

A notable feature of Fig.5-5 is the velocity of flow at and near B, the toe of the dam, which is apparently unbounded and produces a condition favoring piping and dangerous erosion of the foundation materials.

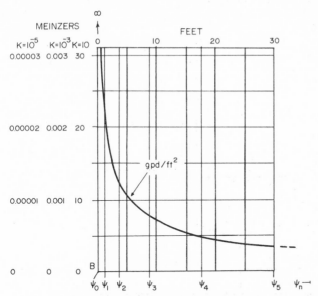

Fig.5-5. Exit flow (discharge) downstream from the dam shown in Fig.5-3.

Dams, especially concrete dams, rarely are constructed with their bases resting on flat surfaces at the same elevation as the floor of the reservoir. Instead, the base is located in an excavated trench. In addition it is common practice to install a cut-off somewhere beneath the dam. In concrete dams the cut-off commonly is a deep narrow trench filled with concrete or a curtain formed by grouting. In earth or rock dams the cut-off generally is a grout curtain, an excavated trench filled with impermeable material, or sheet piling.

The advantages that derive from setting the dam below ground level and installing a cut-off below the dam are suggested by Fig.5-6, which shows a dam on a layer of permeable material which lies above an impermeable basement. The equations for the flowlines and equipotential lines are not known but it is possible to locate the lines fairly accurately from data provided by model experiments and from calculations using approximations.

When the base of the dam is set below ground elevation and a cut-off is constructed, as in Fig.5-6, there is a change in the flow net compared to that in Fig.5-3 that results in the following advantages:

(1) The uplift pressure at the heel of the dam is reduced and the total uplift pressure downstream from the cut-off has been diminished. Accordingly, the moment of uplift forces tending to lift the dam has been reduced.

(2) The danger of piping and erosion at the toe of the dam has been reduced or eliminated.

(3) The longer flow paths along the streamlines below the cut-off causes a notable decrease in the exit velocities downstream in the proximity of the dam and reduces total seepage under the dam.

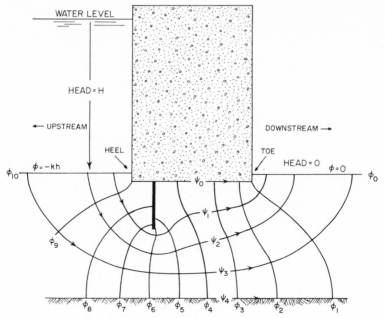

Fig.5-6. Hydrodynamic net beneath a dam with its base below ground level and with an impermeable cut-off near the heel of the dam.

A third method of altering the flow net beneath a dam is indicated in Fig.5-7. Grout pumped into drilled holes sloping upstream from a gallery within the dam provides a more or less impermeable barrier to seepage, and drain holes sloping downstream intercept much or all of the water that passes beneath (or through) the grout curtain. The drain holes greatly reduce the average uplift pressure under the portion of the dam between the toe and the gallery and at the same time practically eliminate the possibility of piping at the toe.

FLOW SURFACES AND EQUIPOTENTIAL SURFACES

Flow nets are two-dimensional and generally are represented in vertical section with the understanding that the flow lines and equipotential lines are the intersections of three-dimensional flow surfaces and equipotential surfaces with the plane of the section. Mathematical analysis of these surfaces is very difficult, if not practically impossible. However, graphical methods assist in the development of approximate solutions of simple, symmetrical problems.

As an example, Fig.5-8 shows an idealized cross-section and plan of a gravity concrete dam constructed in a symmetrical valley. Fig.5-8A indicates the location in vertical section of flow lines and equipotential curves, assuming a uniformly permeable dam foundation of infinite depth. Fig.5-8B shows elevation contours and

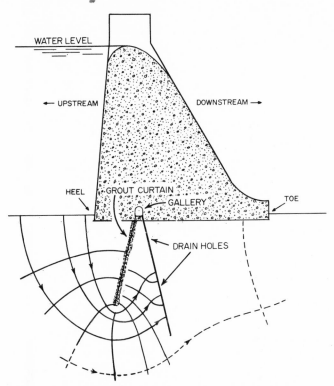

Fig.5-7. Flow net (no scale) beneath a concrete gravity dam with a grout curtain and drainage holes in the foundation.

the intersection of the equipotential surfaces (ϕ surfaces) with the ground surface beneath the dam (stippled area) and the intersection of the flow surfaces (ψ surfaces) with the surface upstream and downstream from the dam. Streamline flow beneath the dam everywhere is assumed to be in vertical planes which are normal to the length of the dam and parallel to the contours.

The intersections of the flow planes with the ground surface downstream from the dams yield curves of equal exit velocities. Thus curve ψ_5 in Fig. 5-8B is the locus of a line traced on the floor and sides of the valley where it should be expected that the quantity of discharge will be constant.

UNCONFINED FLOW THROUGH UNIFORMLY PERMEABLE MEDIA

The first scientist to study and develop a successful formula for unconfined flow of groundwater was Dupuit (1863) His formula with modifications and elaborations by later workers is basic to a great variety of investigations of hydrologic problems such as flow of water into wells and through earth dams.

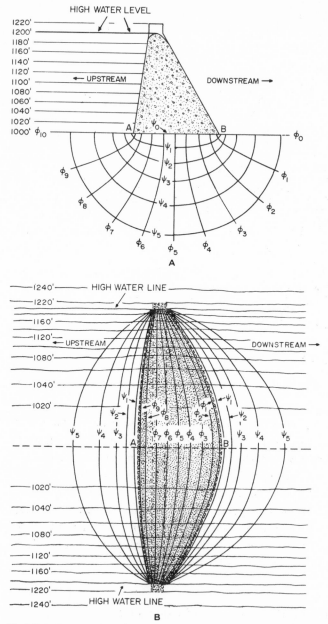

Fig.5-8.A. Cross-section of a concrete gravity dam resting on a uniformly permeable medium of infinite depth. Underground water movement is indicated by flowlines (ψ) and equipotential lines (ϕ). B. Plan of the same dam. Shown are equipotential lines (ϕ) in area beneath dam (stippled) and flow lines (ψ) upstream and downstream from the dam.

Unconfined flow of groundwater is flow in saturated materials at and below the groundwater table in response to head (pressure) differences. The groundwater table is a *free surface* subject to a variety of fluctuations but is always characterized by a pore pressure equal to atmospheric pressure. In common practice a head of zero is assigned to the surface of the groundwater table on the assumption that atmospheric pressure everywhere at the surface is essentially constant.

Dupuit's principle states that the average or mean velocity of seepage (discharge) across a vertical plane perpendicular to the general direction of flow is equal to the coefficient of permeability (seepage coefficient) multiplied by the slope of the free surface at the location of the vertical plane.

Mathematically the principle is stated as:

$$V = - k \, (\mathrm{d}H/\mathrm{d}s) \tag{5-13}$$

in which V is the discharge velocity, H is the head measured from a horizontal datum plane, and s is a measure of distance in the direction of flow. The constant k is the same constant as that used in solutions based on Darcy's law.

For unconfined flow above a horizontal impermeable layer Dupuit's formula, given without proof is:

$$- q = k \, (h_1^2 - h_2) / 2L \tag{5-14}$$

where q is the discharge per unit width through a vertical section parallel to the direction of flow and h_1, h_2, and L are as indicated in Fig.5-9.

Groundwater flow below the water table and above an inclined boundary at the top of an impermeable medium is indicated in Fig.5-10. The angle of inclination of the inclined boundary is α, measured either from the horizontal or from the vertical. The head h_0 at any point is measured vertically from the inclined surface to the free surface consistent with the fact that pressure due to head is a gravitative phenomenon and the acceleration of gravity is in a direction normal to the earth's

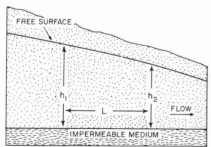
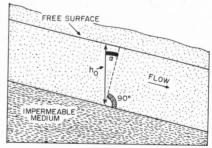

Fig.5-9. Unconfined groundwater flow above a horizontal impermeable surface. See text discussion.

Fig.5-10. Groundwater flow above the surface of an inclined impermeable layer.

surface. For the situation indicated in Fig.5-10 Dupuit's equation assumes the form:

$$q = -kh_0 \tan \alpha \qquad\qquad (5\text{-}15)$$

where q is the quantity of seepage for uniform flow parallel to the impervious boundary.

The groundwater table usually is a curved surface and tends to conform in a general way to surface topographic features (Fig.5-11). If sufficient information

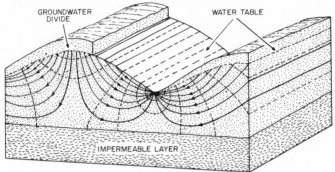

Fig.5-11. Theoretical distribution of flow planes and equipotential planes (dashed lines) in a homogeneous permeable layer above an impermeable layer. Part of the material above the water table is stripped away to show intersections of flow planes and equipotential planes with water table.

from drill holes or geophysical measurements is available, it is possible to prepare a contour map of the water table that resembles a contour map of a portion of the earth's surface. Lines of flow on a contoured water table are always drawn at right angles to the contours.

Because the water table is a free surface it is intersected by both equipotential (equal piezometric pressure) planes and flow planes, as shown in Fig. 5-11. As for confined flow of underground water the streamlines and the planes in which they lie intersect the equipotential planes at right angles.

In general, in regions of topographic relief the slope of the water table tends to be steeper in materials of low permeability than in materials of higher permeability, an observation that is consistent with the fact that under the same head movement of water through slightly permeable materials is considerably slower than through more permeable media.

GROUNDWATER IN SLOPES OF RESERVOIRS

It is emphasized in previous sections of this chapter that investigations prior to construction of a dam should include a careful study of the reservoir site as well as study of the dam foundation. Frequently, the slopes of valleys or canyons are covered with unconsolidated deposits such as glacial till, talus accumulations, or land slides. The purpose of this section is to analyze the effects of groundwater penetration and fluctuation within these deposits and to indicate hazardous situations that might be created in bedrock by filling and drawdown of a reservoir.

Fig.5-12 illustrates successive stages of filling and drawdown in a steep-

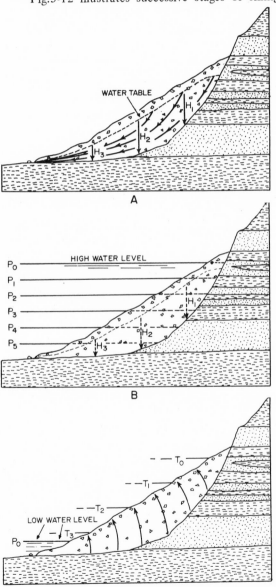

Fig.5-12. Pressure and head adjustments in a talus accumulation during filling and drawdown of a reservoir. See text discussion.

walled reservoir with an accumulation of talus consisting of sandstone and dis-integrated shale (Fig.5-12A). Presumably the material in the talus accumulation, because of its composition, will have a high porosity and a low permeability. The groundwater table can be expected to fluctuate seasonally, depending on weather and the rate of recharge, but fluctuations in the water table are not of particular concern in the present discussions, except to the extent that the head at any point will fluctuate also.

In Fig.5-12A a certain location for the water table is assumed and the different heads at various locations, H_1, H_2, and H_3 are indicated. Also indicated are several streamlines that show the directions of slow movement of groundwater toward the bottom of the talus slope. The movement is of the general kind specified in Dupuit's equation for unconfined flow above an inclined impervious surface.

Fig.5-12B shows the consequences of slow filling and/or long-term storage of water in the reservoir. P_0, P_1, P_2, etc. are hydrostatic pressures in the reservoir, increasing with depth. These pressures will cause slow percolation into the talus accumulation, and, given sufficient time, pore pressures approximately equal to the hydrostatic pressures will develop along essentially horizontal planes (dashed lines, Fig.5-12B). When the pore pressures have stabilized, the whole system reaches a state of equilibrium and movement of water into the talus accumulation ceases. Pore pressures within the talus accumulation, except near its top and just below high water level, considerably exceed pore pressures before filling of the reservoir.

Fig.5-12C shows the consequences of drawdown, especially rapid drawdown, of the reservoir. Precisely what happens depends on the rate of drawdown, sug-gested by T_0, T_1, T_2, etc., which indicate the water level in the reservoir at various times during drawdown, and on the permeability of the talus material.

With instantaneous drawdown the various pressures P_0, P_1, P_2 would be the pressures in horizontal equipotential surfaces, and flow of pore water would be vertically upward across surfaces of diminishing potential (pore pressure). Off-setting the tendency to rise is the normal head in the material before reservoir filling, but this is not sufficient to nullify the pore pressures generated by reservoir filling.

Movement of groundwater under actual conditions of drawdown is very complex and constantly changing. Fig.5-12C shows flowlines of the general con-figuration that can be expected at various stages of drawdown. Movement of pore water from the deeper to the shallower parts of the talus accumulation under notable pressure differentials and the pressures themselves tend to induce instability in the accumulation. A likely consequence is the failure of the slope and collapse into the reservoir, unless the talus materials are moderately to highly permeable and permit rapid adjustments of pressure during drawdown.

In Fig.5-12 the shear stress, τ, within the talus accumulation prior to reservoir filling is insufficient to cause failure along one or multiple slip surfaces and is less than some critical value $\tau_{\text{crit}} = (\sigma_n - P) \tan\Phi$. In this expression σ_n is the normal stress across a potential slip surface tending to prevent slipping because of frictional resistance, P is the pressure exerted by interstitial water and opposing σ_n, and Φ is the angle from the horizontal of a potential slip surface. Filling of the reservoir increases pore pressures in the talus, but these pressures are counterbalanced by the hydrostatic pressures within the reservoir, and the system remains in approximate equilibrium.

However, if, during drawdown, the flow of pore water from the accumulation does not keep pace with diminishing hydrostatic pressures in the reservoir, unbalanced residual pressures are created. The critical stress for failure along slips now is expressed as $\tau_{\text{crit}} = (\sigma_n - P') \tan\Phi$, in which P' is the sum of the initial pore pressure, P, and the residual pressure from reservoir filling. Because $P' > P$, the critical shear stress that existed prior to reservoir filling, has been reduced, and failure along slip surfaces is a likely possibility. That is, addition of residual pressures to initial pressures has in effect reduced the resistance of the talus accumulation to shear failure.

An especially hazardous condition is indicated in Fig.5-13. A slide has

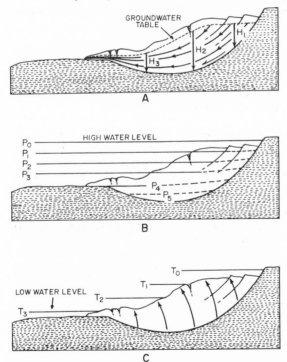

Fig.5-13. Pressure and streamline flow adjustments in an earth slide as a consequence of filling and drawdown of a reservoir.

developed in shale with a typical curved lower surface of shear. Considering the composition of the materials in the slide it is expected that the average permeability is low and the average porosity is high. However, internal dislocations of the materials during development of the slide probably increased the permeability somewhat with respect to the original undisturbed materials, and it is assumed that water from a reservoir has relatively easy, long-term access to the slide materials.

Filling of the reservoir with time will develop pore pressures approximately equal to hydrostatic pressures in the reservoir, and drawdown will generate unbalanced forces that will tend to cause movement of materials in the slide into the reservoir, and, possibly, generation of additional surfaces of slip beyond the existing heel of the slide.

The processes indicated in Fig.5-13 also are capable of producing instability in slopes in bedrock other than shale.. Particularly susceptible to pore-pressure build-up during reservoir filling are highly fractured and/or weathered rocks of all kinds. Failure of slopes on the sides of a reservoir for the reasons outlined above, is especially frequent during drawdown but may not present ummanageable problems when the reservoir is partly or nearly empty. Of more critical concern are rock and earth slides into a reservoir when it is full, and destructive overflow of the dam is a consequence. Hazardous geologic conditions associated with full, or nearly full, reservoirs are discussed in a subsequent chapter (Chapter 6).

REFERENCES

Chugaev, R. R., 1971. Seepage through dams. In: F. T. Chow (Editor), *Advances in Hydroscience.* Academic Press, New York, N.Y., 7: 283-325.

Darcy, H., 1856. *Les fontaines publiques de la Ville de Dijon.* Dalmont, Paris, 674pp.

Davis, S. N. and DeWiest, R. J. M., 1966. *Hydrogeology.* Wiley, New York, N.Y., 463pp.

Domenico, P. A., 1972. *Concepts and Models in Groundwater Hydrology.* McGraw-Hill, New York, N.Y., 405pp.

Dupuit, J., 1863. *Études théoriques et pratiques sur le Mouvement des Eaux dans les Canaux découverts et à Travers perméables.* Dunod, Paris, 2nd ed., 304pp.

Harr, M. E., 1962. *Groundwater and Seepage.* McGraw-Hill, New York, N.Y., 315pp.

Pietraru, V., 1968. Sur la solution des problèmes d'infiltration à l'aide des ordinateurs électroniques. In: *The Use of Analog and Digital Computers in Hydrology*, I. International Association of Scientific Hydrology, pp.175–188.

Chapter 6

GROUNDWATER HYDROLOGY OF ANISOTROPIC MEDIA

INTRODUCTION

Most rocks and accumulations of unconsolidated natural materials are anisotropic; that is, they have properties which are not the same in all directions. In this
chapter attention is given to groundwater hydrology as it is related to the movement of water through anisotropic materials beneath standing or running bodies of
water at the earth's surface and beneath dams.

The emphasis is on groundwater hydrology in valleys because it is there that
most major dams are constructed. The special problems that are encountered in
terrains underlain by soluble rocks are treated separately in Chapter 7.

Throughout this chapter the terms *permeable* and *impermeable* are used
frequently. The reader should understand that the terms are used in a relative sense
and that "permeable materials" are those that are of engineering concern because of
their capacity to transmit groundwater in impermissible amounts. In contrast,
"impermeable materials" have permeabilities so low that they present no problems
in the design and construction of dams.

HYDROLOGY OF ANISOTROPIC UNCONSOLIDATED VALLEY FILL

Contemplated construction of a dam, particularly an earth- or rock-fill dam in
a valley of low to moderate profile, requires careful preliminary assessment of the
lateral and vertical distribution and physical properties of the various kinds of
materials in the unconsolidated valley fill, as well as interpretation of the bedrock
geology. Where valley fill forms a relatively thin mantle on bedrock it is usual
practice to excavate all of the fill beneath the dam, or to excavate a cut-off trench
to bedrock, both on the floor and up the sides of the valley. However, at sites
where valley fill is so deep that estimated costs of excavation to bedrock become
economically prohibitive, consideration must be given to construction of the dam
on unconsolidated foundation materials.

In drill-testing and geophysical surveys of unconsolidated foundation
materials of great depth a hoped-for situation is one in which the valley floor proves
to be underlain with nearly uniform materials with permeabilities and other
physical properties at least equal to those of the materials to be used in construc-

tion of the impermeable portion of the dam. This condition is sometimes attained in valleys in soft fine-grained rocks such as shale, especially where the drainage system upstream is in similar rocks.

More commonly, particularly in valleys occupied by major streams, the valley fill is highly anisotropic, with complexly and seemingly inconsistently distributed layers and irregularly-shaped zones of permeable accumulations alternating both horizontally and vertically with zones of less permeable materials. In steeper-walled valleys the probability exists that products of slope-failure are intermingled with valley-floor deposits from running water or moving ice (see Chapter 3).

Assuming that the bearing strength of the valley fill is adequate to support a heavy structure, the problem in a valley with very deep fill becomes one of locating zones of permeable materials that might result in excessive seepage or even dangerous piping beneath the dam. When these zones are located and appraised, consideration can be given to possible remedial measures such as excavation of a cut-off trench to a predetermined depth, curtain grouting, sheet or column piling, or the placement of an impermeable blanket on the reservoir floor upstream from the dam.

Under many circumstances calculation of an "average" permeability in heterogeneous foundation materials is of doubtful value, and may lead to unwarranted assumptions in dam design and construction that could result in major problems after reservoir filling.

A basic understanding of the groundwater hydrology of valley fill requires an understanding of the processes of excavation of valleys and the processes that result in accumulation of unconsolidated materials on valley floors and valley sides. For a review of these processes the reader is urged to turn to Chapter 3, where attention is given to deposition by running water, moving glacial ice, downslope movement of products of mass wasting, and filling of lake basins in valleys.

In this section emphasis is placed on the groundwater hydrology of anisotropic valley fill, whatever its origin, and especially in unconsolidated materials that locally are moderately to highly permeable. A classification of relative permeabilities of common unconsolidated materials on valley sides and in valley floors is given in Table 6-1.

In many stream-cut valleys of low to moderate profile consideration of the nature of the valley fill at a proposed dam site is primarily directed toward analysis of the depth and internal characteristics of alluvial deposits beneath floodplains. As outlined in Chapter 3, the total history of a valley and the unconsolidated deposits in it may be very complex, but whatever the history of the valley may be, it is certain that there is present some kind of a bedrock floor that was carved out of more or less lithified rocks prior to the deposition of the alluvial deposits beneath the floodplain.

To gain insight into possible vertical and horizontal variations in permeability in floodplain deposits, consideration is given to the processes of accumulation of

TABLE 6-1

Relative permeabilities of unconsolidated materials

I.   Generally low permeabilities
       A. Alluvial clay, mud, and silt accumulations.
       B. Altered volcanic ash (may have notable swelling properties).
       C. Alluvial sand and gravel with compact clay matrix.
       D. Glacial till with compact clay or rock flour matrix.
       E. Buried mud flows.
       F. Landslides, either consisting of clayey components, or of rock fragments in a
          continuous, compact clay matrix.

II.   Generally with moderate permeabilities — sufficient to cause concern as to potential for
       seepage.
       A. Alluvial sand or gravel, poorly sorted, and containing some clay matrix.
       B. Very fine-grained sand.
       C. Unaltered volcanic ash.
       D. Glacial till with a low to moderate content of clay or rock flour.
       E. Landslides consisting of rocky fragments with moderate amounts of clay-size matrix.
          Some voids, not interconnecting, may be present.
       F. Loess.

III.   Generally with high permeabilities
       A. Well-washed and/or sorted alluvial sand or gravel.
       B. Medium- to coarse-grained, angular pyroclastic deposits.
       C. Glacial till with lenses or irregular bodies of stream-transported materials or glacial till
          in which piping has locally removed fine-grained matrix.
       D. Buried sand dunes.
       E. Landslides consisting of angular rock fragments with interconnecting interstitial
          voids.

---

alluvial deposits in a simple valley of low to moderate profile and low to moderate gradient occupied by a stream of inconstant volume and velocity. An assumption is made that the stream has available to it for transport from the upper reaches of the stream and from tributary valleys materials ranging from clay-size to the size of coarse gravel, and that during periods of high water the stream has the capacity to carry particles of all these sizes, as bed load or in suspension.

A valley of simple configuration in plan and profile is shown in Fig.6-1. Valley fill is preponderantly alluvium deposited on the bedrock floor of the valley. The flood plain contains a sinuous stream and exhibits arcuate scars (usually depressions) indicating former locations of stream channels on the flood plain. Although the stream is meandering, and bank-cutting on the outsides of curves is an active process, it is assumed that during periods of major flooding an existing channel may be abandoned and a new one created, so that, at any given time, the stream may occupy a channel anywhere on the flood plain and possesses the capacity to transport bedload and suspended load. This assumption is made to provide a mechanism for efficient aggradation of the floodplain. The materials

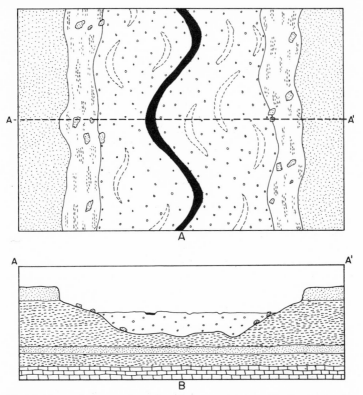

Fig.6-1. Plan and cross-section of a valley containing a wide floodplain. Arcuate areas outlined with dashed lines are abandoned stream channels.

beneath the flood plain in Fig.6-1 are not shown as differentiated in terms of composition and permeability, but it is assumed that variations exist, both verti-cally and horizontally.

Materials in the valley fills are expressions of many variables, including gradients, volume and velocity of streams, and the nature of bedrock and surficial materials upstream from the site of deposition. Accordingly, valley fill may consist entirely of very fine-grained, relatively impermeable clay and/or silt or entirely of porous sand and/or gravel. Usually, a mixture of fine-grained, medium-grained, and coarse-grained materials is supplied to the site of deposition, and it is the sorting and separation of these materials that lead to permeability variations.

Aggradation by a stream of a floodplain such as the one illustrated in Fig.6-1 is indicated in Fig.6-2, and produces heterogeneity in floodplain deposits. When the stream is confined to its channel, bank undercutting results in growth of more or less permeable point-bar accumulations inside curves in the stream channel (Fig.6-2A). In straight stretches symmetrical cross-sections tend to develop, and bank erosion is not significant. During intervals of very high water, when the stream

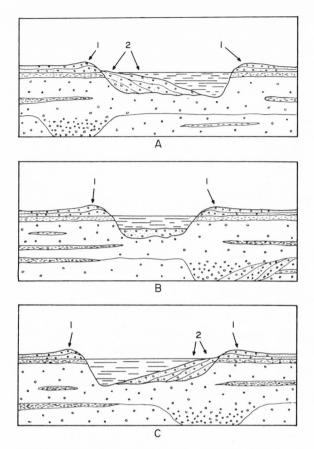

Fig.6-2. Cross-sections of a stream channelway such as the one indicated in Fig.6-1. Point bars (2) develop inside curves (Fig.6-2A and C). When the stream overflows its banks, levees (1) are built up.

overflows its banks, a *levee* develops adjacent to the channelway.

Typically, medium- to coarse-grained materials in the levee give way laterally to less permeable, finer-grained, even clayey materials toward the edges of the floodplain. Materials of bank overflow commonly fill depressions, such as abandoned channelways and pits formed by piping by movement of underground water with fine- to coarse-grained materials, depending on their distance from the channelway and the caliber of the stream load.

Changing conditions during accumulation of deposits of alluvial fill beneath a floodplain may be subtle or gross. Time intervals, during which floodplain aggradation produces extensive layers of fine-grained, relatively impermeable clayey and silty materials, may alternate with intervals during which deposited materials are dominantly medium- to coarse-grained and permeable, as in Fig.6-3. A change from deposition of fine-grained to coarser-grained material commonly is preceded and/or

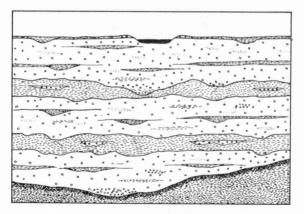

Fig.6-3. Hypothetical cross-section of alluvial fill in a valley in shale. Bank overflow of the stream has built a levee and has formed accumulations of gravel and silt (or clay) on the flood plain. Beneath the floodplain layers of impermeable clay and/or silt alternate with layers of permeable gravels. Filled channels exist in both the permeable and impermeable layers.

accompanied by extensive scouring of the upper surface of a layer of fine-grained material and, in some situations, with partial to nearly complete removal. Thus lateral continuity of a thick or thin impermeable buried layer is not assured.

In the investigation of alluvial fill in a valley prior to construction of a dam consideration must be given to the distribution of permeable zones, not only in a cross-section at a dam site, but also in longitudinal section parallel to the valley. In general, in stream-deposited alluvium, it is reasonable to assume that extensive layers and filled stream channels extend both upstream and downstream for moderate to considerable distances from the dam site. A notable exception to this statement exists in alluvial deposits in lake basins in valleys. Beneath the floodplain of a filled lake basin rapid changes in the distribution of materials of different properties may be spectacular in longitudinal section.

Stages in the filling of a lake basin by alluvial deposits have been reviewed in Chapter 3, to which the reader is referred. In this section, emphasis is placed on longitudinal variations in permeability. Accordingly, sections near the head, in the middle, and near the lower end of a filled lake basin are shown in Fig.6-4 to indicate locations and thicknesses of permeable and impermeable materials.

Fig.6-4A is a section across the upper portion of a filled lake basin into which sediment was transported by a stream. Fine-grained, impermeable materials from a suspended load form a thin layer on the floor of the basin (bottomset beds) and are buried by a relatively thick accumulation of permeable gravels. As lake filling by alluvium proceeds, the thickness of the layer of impermeable material on the lake bottom increases in the direction of the lower end of the lake, and, concomitantly, the thickness of the permeable gravels decreases as in Fig.6-4B and 6-4C.

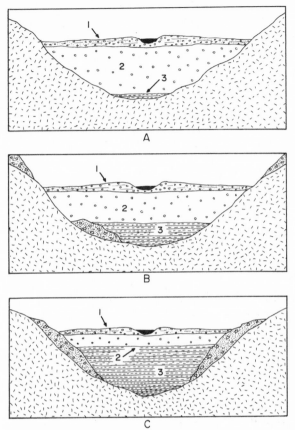

Fig.6-4. Hypothetical cross-section of a filled lake basin in a glaciated valley containing deposits of glacial till. Topset beds (*1*) and foreset beds (*2*) are assumed to be permeable gravels. Bottomset beds (*3*) from fine-grained suspended load are assumed to be essentially impermeable. Thicknesses vary depending on the caliber of the load supplied by the stream.
A. Section near upper end of filled lake basin.
B. Section near middle of basin.
C. Section near outlet end of lake basin.

In valleys with moderate to steep slopes it should be expected that products of mass wasting will be intermingled with the alluvial deposits. In glaciated valleys the additional possibility exists that the alluvial deposits have buried accumulations of more or less permeable glacial till. Three examples of valley fill, in which alluvial deposits and products of slope failure and/or glacial deposition cover the valley floor, are shown in Fig.6-5, and suggest the great number of possible variations in permeability, both laterally and vertically, in valley fill of diverse origins.

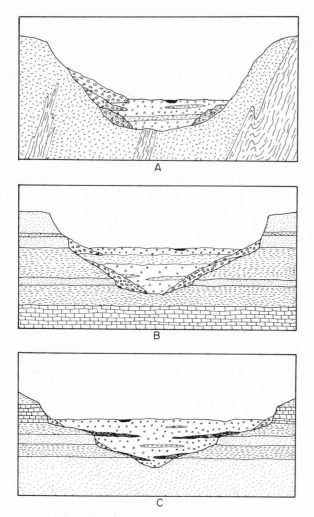

Fig.6-5. Valley fill of diverse origins.
A. Talus from slope failure, glacial till, and stream-deposited alluvium constitute fill in a glaciated valley.
B. Permeable talus accumulations cover slopes beneath layered alluvial deposits.
C. Buried mud flows extend from shale benches into stream deposits.

PROCESSES TENDING TO MODIFY PERMEABILITIES OF UNCONSOLIDATED DEPOSITS

A common experience in an excavation below the water table of sands and gravels of high average permeabilities is discovery that water flows into the excavation are localized by a more or less intricate system of channelways of underground flow. Pumping tests in wells and boreholes to determine average permeabilities may yield inconsistent results that do not appear to be in agreement with the presumed origin and visible physical characteristics of the materials.

In addition to the various controls that determine subtle to gross differences in permeability of originally deposited materials, there are several processes that act to increase or reduce permeabilities after deposition. For example, as in Fig. 6-6,

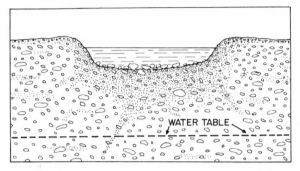

Fig.6-6. "Armored" or "silted" stream channel in initially permeable gravels.

initially permeable gravels beneath and in the banks of a stream may be rendered practically impermeable through invasion and void-filling by silt and clay-size particles derived from the stream. A result is an "armored" or "silted" channel in which the stream may be contained at some distance above the groundwater table.

During bank overflow gravels in the floodplain on both sides of a stream channel may be penetrated by fine-grained materials to depths of a few inches to several feet so as to notably reduce initial permeability. Growth of vegetation and soil development likewise tend to reduce the permeability of materials exposed at the surface of the floodplain.

Movement of groundwater through permeable zones below the floodplain causes progressive changes in permeability, especially in accumulations that are initially moderately to highly permeable. Even though the average permeability may be high, it should be expected that differences exist both horizontally and vertically because of anisotropism related to the mechanisms of deposition and that the velocities of groundwater movement will be greater through the more permeable zones than through less permeable zones. By a process similar to that operating to produce armored stream channels fine particles are swept out of the interstices of the more permeable materials by piping and carried into zones of lower ground-water velocity. This process leads to the development of "armored", highly localized underground channels of flow that, when intersected in excavations, prove to be the main channelways of underground water circulation. For this process to be effective it must be assumed that groundwater has easy access to and egress from permeable accumulations.

In testing permeabilities of alluvial deposits by pumping from wells or by forcing water into drill holes under controlled conditions it should be anticipated that permeabilities within armored channelways ordinarily will exceed those outside the channelways perhaps by several orders of magnitude. As a consequence of dam construction on or within alluvium, changes in groundwater flow patterns beneath the dam during reservoir filling may promote slow to rapid growth of armored channelways that did not previously exist.

Other processes that contribute to changes in permeability of alluvial materials after initial deposition include interstitial deposition of chemical compounds, especially iron and manganese oxides, solution of particles of water-soluble substances, such as limestone and dolomite, and compaction by the load of overlying layers.

PERMEABILITY AND SEEPAGE IN UNCONSOLIDATED DEPOSITS

Except under the most favorable circumstances, precise, quantitative estimation of the potential flow of groundwater through unconsolidated deposits beneath a dam is very difficult, if not impossible, and recourse must be made to estimates of orders of magnitude, calculations based on empirical relationships derived from experience, and model experiments. To obtain reasonable values for permeabilities materials beneath the dam must be carefully investigated by drilling, pressure and pumping tests, laboratory examination of samples, and, where indicated, geophysical measurements. Depending on the configuration of the base of a dam and the space distribution of materials of different permeabilities beneath, upstream, and downstream from the dam, flow patterns range from simple to very complex.

In the usual situation both vertical and horizontal anisotropism in natural materials prevent consistent determination of permeabilities, and the best that can be expected is an estimation of the average permeability from a large number of measurements. Accordingly, it can not be assumed that estimations of quantities of flow will be any more accurate than the average permeabilities obtained from field and laboratory measurements.

Fig.6-7 indicates two kinds of flow (seepage) beneath a dam. Both are confined flow but with totally different configurations of flow lines and equipotential lines (dashed). In Fig.6-7A the flow is bounded upstream by an equipotential surface at the floor of the reservoir where the potential is equal to H, the head exerted by the water in the reservoir, and downstream by a surface of zero head. In Fig.6-7B the flow is bounded above and below by parallel impermeable layers, and the lines of equal head are perpendicular to the confining surfaces and to the streamlines.

An analysis of the kind of confined flow suggested by Fig.6-7A is made with

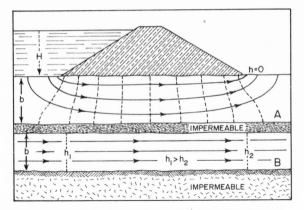

Fig.6-7. Two kinds of groundwater flow beneath a dam.
A. Confined flow above an impermeable layer.
B. Confined flow between parallel impermeable layers.

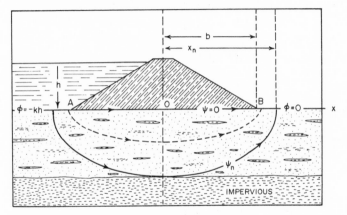

Fig.6-8. Seepage beneath a dam. See text discussion.

the aid of Fig.6-8, and suggests the kinds of problems that are encountered in more complex situations. It is assumed that an average permeability has been ascertained by field measurements, and that, although the materials beneath the dam are actually inhomogeneous, they can be considered to be homogeneous in calculations of the order of magnitude of seepage.

The boundary conditions in Fig.6-8 are of the simplest kind. A dam with a flat base rests on a layer of permeable material above an impermeable layer. A horizontal equipotential surface along the floor of the reservoir has a value of $\phi = kh$, in which k is the average permeability of the permeable layer, and h is the head in the reservoir. Downstream from the dam the head is equal to zero, and $\phi = 0$. Flowlines (streamlines) at the base of the dam ($\psi = 0$) and tangent to the impervious layer (ψ_n) have equations such that the total quantity of flow between them (Q) can be calculated from the relationship:

$$Q = (kh/\pi)\cosh^{-1} (X_n/b) \tag{6-1}$$

in which X_n and b are the dimensions indicated in Fig.6-8.

Now, let $X_n/b = u$. Expanding $\cosh^{-1}$, a hyperbolic function, we obtain:

$$Q = (kh/\pi) [\log_e 2u - (1/4u^2) - (3/32u^4) - (15/288u^6) - \ldots] \tag{6-2}$$

In Fig.6-7B the flow is "pipeline" flow in which Darcy's equations in their simplest expression are applicable. In Fig.6-7B, according to Darcy's law:

$$Q = KA(dh/dL) \tag{6-3}$$

in which Q is the quantity of water flowing through a cross-section of area A in a medium with a coefficient of permeability K and a hydraulic gradient of dh/dL.

For an aquifer of thickness b and width W the quantity of water flowing through a cross-section of the aquifer is:

$$Q = KbW(dh/dL) \tag{6-4}$$

Q is obtained in practice by drill hole measurements of b and W, field or laboratory measurements of K, and measurements of head in bore holes at points separated by a distance L.

Then:

$$Q = K (bW)(h_1 - h_2)/L \tag{6-5}$$

If we designate $T = Kb$ as the *coefficient of transmissibility*, then:

$$Q = TW(h_1 - h_2)/L \tag{6-6}$$

HYDROLOGY OF ANISOTROPIC BEDROCKS

Permeability in bedrocks is of two kinds. *Primary permeability* is determined by the mode of origin of a rock and includes features such as bedding (stratification) in sedimentary rocks, layering in pyroclastic rocks, and layering (foliation) in met-amorphic rocks. In general, groundwater flow is greater in directions of layering than across it, and, depending on the origin of the rock, ranges from essentially zero to flow of a large magnitude. *Secondary permeability* is superimposed on a rock by a variety of natural processes including dislocations by crustal movements, by weathering, and by chemical alterations associated with solutions of deep-seated origin. (See Chapter 4.)

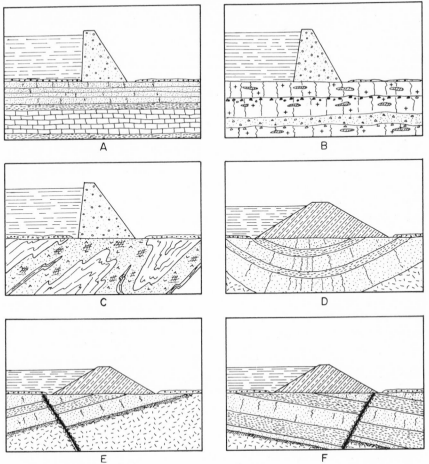

Fig.6-9. Idealized cross-sections of dams showing various kinds of zones of potential seepage in bedrock. (Cont'd. on p. 146.)

A. Brittle, fractured sandstones in horizontal sedimentary sequence beneath dam present a potential for seepage.

B. Dam is situated on basaltic lava flows and interlayered pyroclastic deposits. Lava flows are fractured, brecciated at their tops, and contain lava caves.

C. A brittle layer of quartzite in tightly folded metamorphic rocks is likely to contain numerous intersecting fractures.

D. Sandstone layers alternating with shale layers in a syncline contain fractures associated with development of a syncline.

E. A fault zone provides access of water to brittle sandstone layers which dip upstream.

F. A fault provides egress for water moving through inclined, brittle sandstone layers.

Moderate to high primary permeabilities exist in relatively few rocks. Notable exceptions are found in some medium- to coarse-grained clastic and pyroclastic rocks with minor amounts of interstitial cement or matrix and in lava flows with breccia tops or containing lava caves or numerous gas cavities. Crystalline igneous

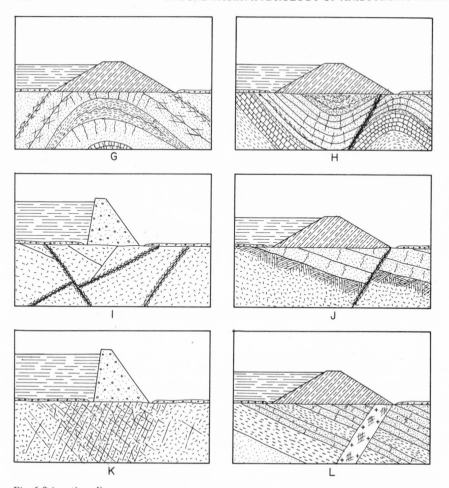

Fig. 6-9 (continued).
G. Brittle sandstone layers have been extensively fractured during development of an anticline.
H. Fractured sandstones in folded rocks are intersected by a fault zone which expedites groundwater movement to the surface beneath the dam.
I. Faults in brittle crystalline rock provide channelways for groundwater circulation.
J. Fractured sandstones, a weathered zone on granite beneath the sediments, and a fault zone create channelways for subsurface circulation of water.
K. Extensively jointed crystalline rocks are permeable to groundwater flow.
L. A closely jointed igneous dike intersects a sedimentary sequence and provides a channel for groundwater movement.

and metamorphic rocks, partly to completely crystallized sedimentary rocks, and clastic rocks with grain interstices filled with cement or clay matrix generally are esssentially impermeable.

Interconnecting large to small channels of flow associated with secondary permeability are of great concern in investigations of dam and reservoir sites, and

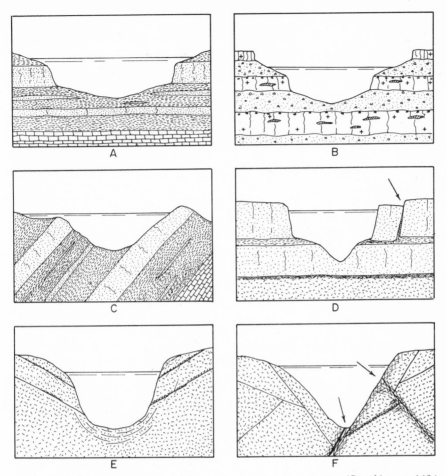

Fig.6-10. Idealized cross-sections of valleys at dam and/or reservoir sites. (Cont.'d. on p. 148.)

A. Jointed sandstones present potential for seepage around dam abutments when reservoir is full.

B. Basaltic lava flows and layers of pyroclastic rocks create potential for seepage. Lava flows are jointed, have breccia tops, and contain lava caves.

C. Fractured sandstones in a strike valley are prone to seepage.

D. A subsided block (arrow) has created an open channelway in a massive horizontal sandstone layer.

E. Gravity-slip faults and fractures formed by elastic rebound produce potential zones for groundwater movement in a glaciated valley.

F. A strong fault system renders crystalline rocks permeable on one side of a valley.

every effort should be made to locate them and assess their potential for leakage through the abutments or beneath a dam.

Fractures that have formed in hard, brittle rocks, including many metamorphic rocks, most igneous rocks, and well-cemented clastic sedimentary rocks, generally are more continuous and more likely to be open to groundwater pene-

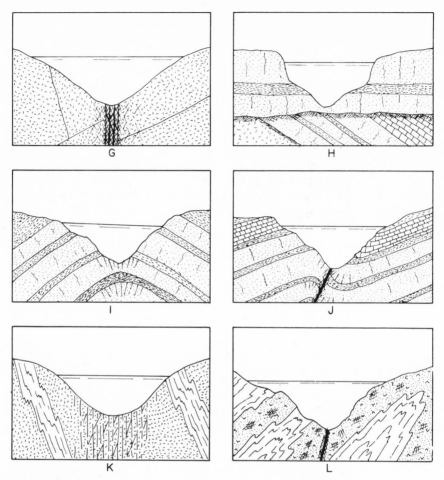

Fig.6-10 (continued).

G. A wide fault zone promotes deep circulation of water beneath dam.

H. Fractured sandstones and a weathered zone beneath an angular unconformity enable easy circulation of groundwater.

I. Fractured sandstones in an anticline create a permeable zone parallel to the valley.

J. Folded, jointed rocks and a strong fault create a potential for groundwater flow.

K. A sheeted, jointed zone in crystalline rocks creates a permeable zone.

L. Joints in a brittle quartzite layer and a fault produce channelways for underground water circulation.

tration than fractures in soft rocks such as mudstone, shale, and medium- to coarse-grained clastic rocks containing a clay matrix.

Almost an infinite number of possibilities exists with respect to the magnitude and space distribution of zones of potential seepage in bedrock in the vicinities of dams and in the reservoirs behind the dams. A few of the many possible configurations of zones of potential seepage are shown in the diagrams in Fig.6-9 and 6-10.

Secondary permeability developed in soluble rocks such as limestone is considered elsewhere (Chapter 7).

HYDROLOGY OF FILLED RESERVOIRS

In earlier chapters attention was given to potential hazards in reservoirs related to unstable slope configurations and to rapid reservoir drawdown (Chapter 5). Failure of slopes on the sides of reservoirs are especially frequent during drawdown and may not produce unmanageable problems, especially when the reservoir is nearly empty. Of generally more critical concern are rock and earth slides into a reservoir when it is full, and suddenly destructive overflow of the dam is a consequence.

Filling of a reservoir causes adjustments in the groundwater table in adjacent materials. When the reservoir is full for an extended period of time, a groundwater table is established which, at its lowest elevation, coincides with the elevation of the water surface. Seepage from surface recharge tends to build up a free surface below which the materials are saturated and which slopes toward the reservoir. It can be assumed that the configuration of the surface changes with seasonal fluctuations in recharge rates, so that a dynamic system is created in which discharge velocities and pore pressures below the surface (the groundwater table) also fluctuate in materials that were well-drained prior to reservoir filling.

Because the groundwater table is a free surface in contact through unfilled pore spaces with the atmosphere, changes in atmospheric pressure are accompanied by changes in pore pressure in the saturated zone, so that the dynamics of the system are further complicated. An additional factor promoting instability in slopes above the perimeter of a reservoir is wave action which undercuts slopes and oversteepens them.

Any one or a combination of the processes outlined above can create disastrous collapse of the sides of reservoirs involving masses of rock or earth of large to small dimensions. Two examples, which, upon examination, suggest many more are illustrated in Fig.6-11.

In Fig.6-11A a reservoir is located in an extensively glaciated valley in crystalline rocks. Lateral moraines, consisting of a jumbled mixture of large and small bowlders, gravel, and rock flour have been deposited by the glacier that occupied the valley high on its sides. Filling of the reservoir causes an elevation of the water table in the till in the moraines, and because of the cross-sectional configuration of the valley, considerably increases the possibility for sudden downslope movement of the morainal material. Failure of the slopes might occur at any time, that is, during reservoir filling, when the reservoir is full, or during drawdown.

Fig. 6-11B shows the bedrock profile of a canyon eroded by a stream in a

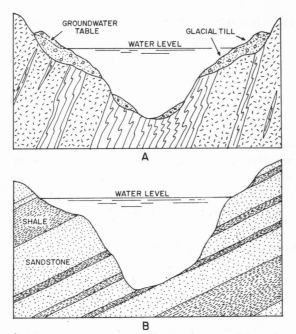

Fig.6-11. Natural conditions promoting failure of sides of reservoirs above high-water elevation.
A. Glacial till deposits in a glaciated canyon in crystalline rocks.
B. Tilted alternation of sandstone and shale.

tilted succession of sandstones and clay-rich shales. Unconsolidated materials in rock slides and stream deposits are not shown. Water from the reservoir, by seepage through the sandstones, comes into contact with the shale layers for a considerable distance into the canyon walls, and by slow penetration of the shales, greatly reduces their strength. Under these circumstances, a highly unstable condition is created, especially where the sedimentary layers dip into the reservoir.

Chapter 7

GEOLOGY AND GROUNDWATER HYDROLOGY OF SOLUBLE ROCKS

INTRODUCTION

Engineering problems of great complexity may be encountered during construction of dams and reservoirs in areas where bedrocks have been more or less dissolved by circulating underground waters. The most spectacular and widespread end result of underground dissolution by groundwater is the development of karst topography and small to large limestone caverns, galleries, and underground rivers. However, even minor solutional features of localized distribution may be of great engineering concern where they contribute to the incompetency and permeability of foundation rocks beneath dams and in the floors and walls of reservoirs. Although solutional activity is known to have produced cavities in many kinds of rocks, by far the greatest numbers of solution openings are in sedimentary limestones and dolomites. In some areas gypsum plays the same role as the carbonate rocks in localizing creation of underground openings by groundwater.

Limestones consist mainly of calcite ($CaCO_3$) and are preponderantly of marine origin, although fresh-water limestones locally are important. The components of limestones and the environments of their deposition are diverse, and grain sizes range from microcrystalline to megacrystalline. Common components are lime mud or ooze, coarser grained clear "sparry" calcite, fossil remains of organisms, more or less rounded, and grains, fragments, or pellets of local origin, or transported by currents to the site of deposition. The compactness and permeability of limestones differ greatly depending on their origin and the extent to which they have recrystallized or have developed interstitial cement. At one extreme of compactness and permeability is practically incoherent porous chalk and shell limestone and at the other is dense, lithified limestone ooze (micrite) and completely recrystallized limestone.

Dolomite, also called *dolostone*, consists largely of $CaMg(CO_3)_2$ and is a close second to limestone in the development of solution openings. Some dolomite formed by original deposition, but most laterally and vertically extensive deposits appear to have resulted from dolomitization of limestones by reaction with saline waters shortly after deposition, by reactions with low-temperature groundwaters after burial, or by reactions with hydrothermal solutions. Commonly, dolomitization results in an increase in the crystallinity and permeability of the replaced limestone.

Gypsum ($CaSO_4$ $2H_2O$) is deposited in sedimentary layers by evaporation of highly saline waters in semiclosed marine basins. Salt, limestone, and dolomite, together with clastic clayey or sandy rocks, commonly are associated with gypsum in so-called "evaporite sequences". Deep burial causes dehydration of gypsum to anhydrite ($CaSO_4$). Subsequent exposure to surface waters results in reconversion of anhydrite to gypsum, generally with a notable volume increase (over 60%). Many surface and near-surface deposits of gypsum grade into anhydrite at depth. Solution cavities are almost unknown in anhydrite but are common in gypsum.

GEOHYDROLOGY OF INITIALLY HOMOGENEOUS SOLUBLE ROCKS

Common rocks that are notably soluble in water are limestone, dolomite, gypsum, and rock salt. Except in exceedingly arid regions rock salt does not exist near the earth's surface because of its extreme solubility, and it is of minor concern in the present context except, possibly, where deep circulation of groundwater is initiated by filling of a reservoir above an evaporite sequence. Limestone and dolomite are somewhat soluble in pure water. At 16°C a liter of pure water can dissolve about 13 mg of pure limestone ($CaCO_3$) and about a fifth of this amount of dolomite – $CaMg(CO_3)_2$. Addition of carbon dioxide from the atmosphere, from decayed organic material, or from sources deep in the earth, notably increases the solubility.

The chief source of carbon dioxide in groundwater is the atmosphere which, under normal conditions, contains about 2.5–3.5 parts per 10,000 (2.5–3.5 × 10^{-4}). The solubility of carbon dioxide in water is a function of temperature, pressure, and acidity or alkalinity. Increase in temperature decreases carbon dioxide solubility, but increase in pressure is accompanied by an increase in solubility, so that calcite and dolomite are notably more soluble in carbon dioxide bearing cold water under pressure than in warm water at low (atmospheric) pressure.

Analyses of surface waters on limestones undergoing dissolution vary widely, depending on local conditions, such as climate and vegetation. A range of values is about 25–150 mg/l of dissolved calcium carbonate, which may be compared with a solubility of about 13 mg/l in pure water. Springs issuing from limestone terranes in areas of temperate climate have concentrations of dissolved calcium carbonate of as much as 300 mg/l.

Gypsum is much more soluble in water than carbonate rocks and has a solubility which does not depend on the presence of carbon dioxide. At 18°C a liter of pure water dissolves 2.6 g of gypsum. The presence of dissolved salt notably increases the solubility of gypsum, so that brine-charged waters, especially from evaporite sequences, are highly effective in causing underground dissolution of gypsum.

The original porosity and permeability of limestone and dolomite layers vary widely, depending on grain size and the degree to which pore spaces are filled with cement or matrix and the extent of recrystallization. Gypsum generally has a low porosity and permeability. Although in rocks such as chalk and shell limestone, initial porosity and permeability play an important role in the circulation of groundwater, most soluble rocks, including limestone, dolomite, and gypsum, are most effectively dissolved by groundwater circulating along interconnecting channels localized along fractures and bedding planes.

The physical and chemical controls of underground solution of soluble rocks range from relatively simple to very complex and depend on climate, topographic relief, geologic structure, composition of the soluble rocks, and primary and secondary porosity and permeability. Soluble rocks are dissolved at the earth's surface in contact with the atmosphere, in the zone of descending groundwater above the groundwater table, and by groundwater in motion beneath the groundwater table. The emphasis is on groundwater in motion, usually of surface derivation, which penetrates soluble rocks and transports the dissolved materials out of the environment. Stagnant underground water becomes saturated with dissolved materials and neither dissolves additional substances or precipitates them.

Dissolution of calcareous rocks as it is related to a changing groundwater table is indicated in a highly diagrammatic manner in Fig.7-1 and 7-2. In both examples it is assumed as a starting condition that a physically and chemically homogeneous layer of limestone with uniform initial permeability overlies a layer of impermeable shale. In Fig.7-1 the starting condition also assumes a valley that retains its depth and cross-sectional configuration throughout a long period of time. In Fig. 7-2 it is assumed that the valley is actively cutting downward during solutional activity in the limestone. In an actual example the processes depicted in Fig.7-1 and 7-2 can be expected to be active simultaneously, thereby increasing the complexity of the total process.

In Fig.7-1A, because of initial low permeability, the groundwater table slopes steeply toward the valley. Flowlines from the groundwater surface show the paths of water flow through the limestone to the stream in the bottom of the valley. The discharge velocity along each of the flowlines according to Dupuit's formula (Chapter 5) is directly proportional to the head and inversely proportional to the distance of travel along each flowline, assuming a constant permeability. Thus, the discharge velocity is less along the deeper flowlines than along the shallower flow lines and solutional activity decreases as a function of depth (indicated by dashed flowlines and sizes of solution openings). A net result is a marked, gradual increase in the permeability of the limestone above the groundwater table and in the limestone immediately below it and diminishing development of permeability at greater depths. Thus, an initially homogeneous limestone layer becomes increasingly nonhomogeneous in terms of the vertical distribution of permeability.

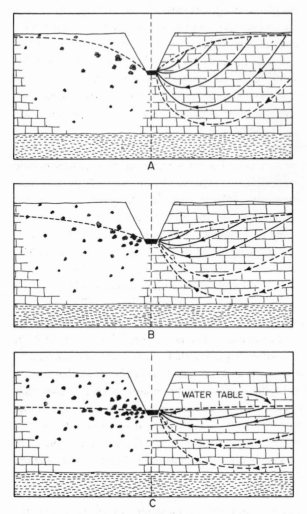

Fig.7-1. Progressive dissolution of limestone with time, as related to changing configurations of water table and flowlines. Downcutting of the valley during various stages of dissolution is assumed to be negligible. Dashed flowlines indicate low to very low discharge velocities of underground water according to Dupuit's principle.

Concomitant with development of permeability in the upper portion of the limestone layer is flattening of the groundwater table as shown in Fig.7-1B and C, and a situation is ultimately reached in which the groundwater table is essentially horizontal and lies below and immediately above extensively dissolved and highly permeable rocks. Secondary solutional permeability extends to the bottom of the limestone layer, but is more related to solution by deep movement of groundwater in the initial stages of development of permeability (Fig.7-1A) beneath a steeply inclined groundwater table than to solution at a time when the groundwater table is

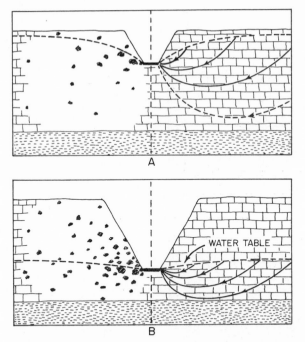

Fig.7-2. Progressive dissolution of limestone with time, as related to changing configurations of water table and flowlines. Downcutting of valley is assumed to be active during various stages of dissolution. Dashed flowline (A) indicates small to negligible discharge velocity of groundwater in agreement with Dupuit's principle.

relatively flat, and deep circulation is essentially insignificant (Fig. 7-1B and C).

Fig.7-2 shows changing permeability by dissolution as a valley is excavated in a limestone layer. Increasing permeability with time above and below a lowering and flattening groundwater table results in extensive solutional activity throughout the limestone layer. Unlike the situation illustrated in Fig.7-1 deep solution increases with time because flowlines are more and more narrowly confined between the groundwater table and an impermeable layer as the valley is deepened.

Various geohydrologic controls of underground water circulation in initially homogeneous soluble rocks, especially limestone and dolomite, are shown in Fig. 7-3. The reader will understand that the flowlines and the groundwater table undergo drastic changes in location and configurations as solution progresses (see discussion of Fig.7-1 and 7-2). Moreover, no attempt is made in the diagrams to indicate the considerable changes in the flow patterns as open, connected fissures, caverns, tunnels, and shafts develop locally. Instead, the diagrams indicate only the locations of zones of minimum, intermediate, and maximum solution of limestone as related to the locations and lengths of flowlines. Maximum solution can be expected to occur where the flowlines are short in length and/or are bunched.

The existence of soluble bedrocks in the vicinity of a proposed dam and

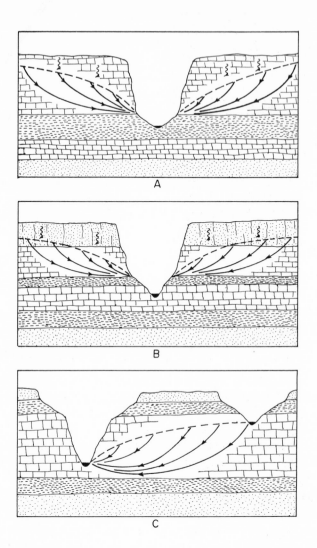

Fig.7-3. Idealized flow patterns indicated by flowlines in limestone. Water table is indicated by dashed lines. Shale sections assumed to be impermeable. For changes in flowlines as related to changing water table, see Fig.7-1 and 7-2 and accompanying text discussion.

A. Limestone exposed at surface is underlain by impermeable shale.

B. An upper limestone is supplied with water by seepage through a jointed sandstone. A lower limestone is not accessible to groundwater of surface origin, and is not affected.

C. A system of circulation is set up between a higher stream and a lower one. Ultimately, the flow in the higher stream may be completely diverted underground by sinkholes to the lower stream.

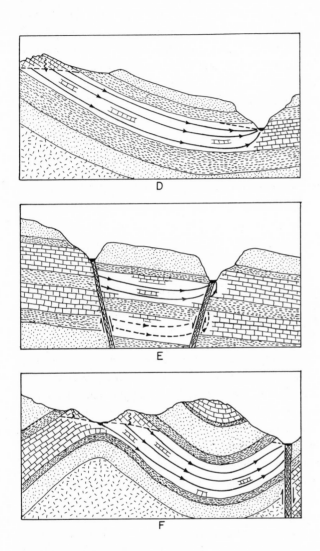

D. Surface water gains access to a limestone exposure and circulates downward to an exit in a stream-cut valley.

E. A tilted, faulted sequence promotes limestone dissolution. Possible circulation in a lower limestone layer is indicated by dashed flowlines. The higher stream ultimately might be entirely diverted by underground flow to the lower stream.

F. Circulation pattern is determined by folding and faulting in a sedimentary sequence resting on crystalline rocks.

reservoir site is cause for considerable concern, and every effort should be made to determine whether groundwaters have penetrated the rocks to produce interconnecting solution openings. All gradations exist between soluble rocks in which groundwater circulation has increased secondary permeability only locally by solution confined to fractures and bedding planes, and rocks in which solution on a massive scale has produced an interconnecting honeycomb of small to large shafts, caves, and galleries.

A first clue to the probable existence of secondary permeability in soluble rocks is obtained by careful examination of surface exposures of the soluble rocks, and in many regions interpretation of surface topographic features that characteristically develop above and within layers of soluble rocks. However, there are many circumstances under which solution cavities develop in buried rocks where there is no indication at the earth's surface that the rocks exist at depth.

DISSOLUTION OPENINGS IN ANISOTROPIC SOLUBLE ROCKS

In previous sections of this chapter the geohydrology of initially homogeneous rocks was discussed from a theoretical point of view. Now, we turn to a consideration of actual natural conditions that are associated with and determine the locations of underground solution openings.

Compositionally and texturally all gradations exist between pure limestone, dolomite, or gypsum and other sedimentary rocks, such as shale, siltstone, and sandstone. Moreover, many limestones are more or less dolomitic and, commonly, the dolomite is confined to certain layers or is localized in irregular gradational bodies within a larger limestone body. Notable vertical and lateral variations in composition and texture in layered soluble rocks are the rule rather than the exception. In addition, many marine limestones vary greatly in thickness where thick, porous reef accumulations give way laterally to thinner accumulations of limy ooze and calcareous remains of small organisms.

Initial permeabilities of soluble rocks display a wide range suggested by the extremely high permeability of poorly cemented, loose accumulations of shells of organisms (shell limestone and chalk) contrasted with the very low permeability of dense, consolidated limestone ooze (micrite) and completely recrystallized limestones. In company with compositional and textural variations, vertical and lateral variations in initial permeability are notable in many examples of soluble rocks.

Although original permeability determined by the mode of origin of a soluble rock locally controls underground movement of water and locations of underground solution openings, by far the greatest number of underground solution openings are related in a complex manner, not only to lateral and vertical variations in composition, but also to the presence or absence of planar or linear elements of structure within the soluble rocks. Outstandingly important in controlling the flow of underground water are bedding planes, joints developed within layers during burial and compaction, and fractures, either faults or joints, superimposed on the rocks by minor to major crustal dislocations.

An optimum condition for initial development of connected solution openings is provided in a nearly pure (almost entirely calcite) limestone containing permeable bedding plane discontinuities and intersected by numerous fractures of any genesis whatsoever. Although bedding planes and fractures are the favored sites for underground circulation of water, it should be expected that the permeability within these planar structures is not everywhere the same, so that initial penetration by water and subsequent solution of adjacent rock will become more and more channelized with time.

Additional channelization, either vertically or laterally, is promoted along intersections of one set of fractures with another or along intersections of fractures with bedding planes.

By giving consideration to the above factors it is seen that the controls that determine the flow paths of underground water and that determine the ultimate precise locations, sizes, and shapes of underground solution openings commonly cannot be identified by geologic studies, partly attributable to the fact that initial subtle or gross differences in the underground environment are no longer apparent because of their elimination by solutional activity.

Considering the complexity of the controls that determine the pattern of development of the openings, prediction of their presence or absence, and an assessment of their locations, sizes, and configurations presents one of the most difficult challenges in the geotechnical sciences.

LANDFORMS IN SOLUBLE ROCKS

In field studies of locations for proposed dams and reservoirs a first indication of the probable existence of solution openings in rocks beneath the surface may be obtained by the identification of landforms that characteristically are associated with soluble rocks.

In many regions of the world, both in plains and mountainous areas, solution of underlying rocks has produced a characteristic topography identified as *karst* (Jennings, 1971; Herak and Stringfield, 1972). Karst is a general term for terrain in which surface and subsurface features, especially the relief and the drainage are directly related to the existence of bedrocks that are notably soluble in natural waters. Typically karst at the earth's surface is marked by intermittent stream flows, dry valleys without streams, and a chaotic, seemingly patternless distribution of landforms. Interrupted valleys and numerous closed basins, some containing lakes, are common. Karst is developing in some areas of the world today, but the processes of karst formation have been active intermittently throughout much of geologic history and reached maxima and minima with changing geologic and climatic conditions. Karst development, *karstification*, as might be expected, is especially rapid in wet warm to tropical regions and slow to dormant in cold dry regions or in hot desert areas.

The development of karst landforms depends on a balance between corrasion by wind, water, and ice and corrosion by surface and underground water. Valley incision in the initial stages of karstification, before an extensive underground system of circulation is established, is similar to valley cutting in relatively insoluble rocks. However, unlike streams in insoluble rocks, much of the transported load is dissolved, and a combination of erosion by the suspended and traction loads and solution of rocks in the stream bed tends to develop steep-walled gorges by rapid downward incision. Preservation of steep walls is enhanced by notable strength in many limestones and dolomites which are capable of sustaining steep to vertical slopes for protracted intervals of time.

If limestone and/or dolomite exist below the bottom of a valley and deep solutional activity is concomitant with valley deepening, a stage may be reached where much or all of the stream flow is diverted to underground channelways and dry or blind valleys develop. In such valleys the corrasion and corrosion formerly confined to a valley are transferred underground and promote the growth of underground channels, some of impressive size and considerable lateral extent. Under these circumstances, minor and tributary streams tend to disappear and only major trunk streams persist, fed in part by surface and underground springs associated with underground flows from areas where streams have disappeared.

The terminology of karst is based on international agreement (Jennings, 1971; Herak and Stringfield, 1972). A distinction is made between karst developed where soluble rocks are exposed at the earth's surface ("bare karst"), or are buried with a thin layer of unconsolidated materials ("covered karst"), and karst consisting of solution openings in soluble rocks beneath moderate to considerable thicknesses of overlying rocks ("subjacent karst"). In areas where surface karst features are observed there is always the probability that subterranean solution openings of more or less complexity exist to a depth limited only by the depth of the bottom surface of a soluble layer. Several terms used to describe land forms associated with karst are defined in following paragraphs.

Dolines. Dolines, also called *sinkholes* and *swallow holes,* are the commonest and most easily recognized surface topographic forms identified with karst. Typically dolines are simple, closed depressions circular to oval in plan and dish- or bowl-shaped, conical, or cylindrical in cross-section. Depending on the elevation of groundwater table and the absence or presence of deeper channels of circulation dolines may be dry or may contain small lakes. As individual dolines in a cluster

expand laterally they merge to produce more complex topographic depressions. A
succession of dolines along a linear trend generally indicates development along a
fracture zone in bedrock or within a tilted layer of soluble bedrock.

Dolines have multiple characteristics and origins, but a broad distinction can

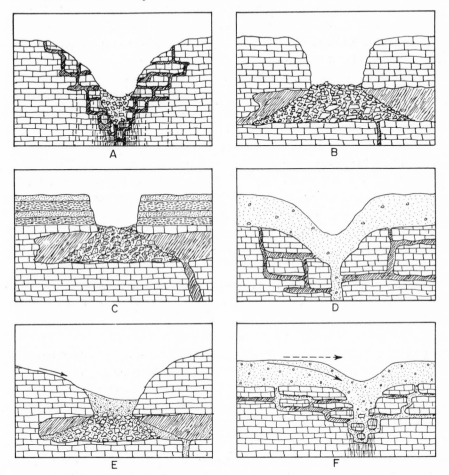

Fig.7-4. Various aspects of dolines.
A. Solution doline formed in a fractured zone in limestone.
B. Collapse doline developed by caving of limestone into a subterranean solution opening.
C. Collapse doline in insoluble sandstone and shale formed by caving into a solution cavity in
limestone.
D. A doline has developed in unconsolidated overburden by settling into a buried solution
doline.
E. A stream has deposited alluvium in a collapse doline.
F. A shallow doline has developed in stream gravel on an alluvial plain. During high water the
stream follows its normal course in a valley (dashed arrow), but during low water stream flow is
entirely diverted to solution cavities (solid arrow) in bedrock.

be made between *solution dolines* which are depressions formed dominantly by solution of bedrock, commonly in a fracture zone, and *collapse dolines,* which form by collapse of surface and near-surface materials into a subterranean opening such as a gallery or cave. Some features associated with dolines of different histories are shown in Fig.7-4.

Jamas. Deep, open pits which connect the earth's surface with underground cavities.

Kamenice. Small, shallow dish-like depressions dissolved by surface water in massive, unjointed limestone or gypsum.

Karren. Minor solution features observed on exposed surfaces of soluble rocks. Shapes of karren reflect subtle or gross differences in solubility, presence or absence of joints, etc., and assume many forms. Sometimes referred to as *solution sculpture.*

Polje. Large steep-sided closed depressions with flat alluvial floors across which streams flow. During periods of high water lakes may form on a polje floor.

Ponors. Also called *swallow holes.* Solution or collapse openings which divert surface water to underground openings. Commonly of small diameter.

Uvala. A large closed depression formed by the coalescence of several adjacent solution or collapse dolines.

Streams in soluble bedrocks commonly develop valleys simultaneously with extensive solutional activity of bedrock. While surface karst features appear, underground movement of water creates interconnecting caves and galleries, and shafts which, with time, are capable of carrying large volumes of water in underground rivers.

During some stage in valley incision in karst areas surface streams may be diverted underground, totally or in part. A consequence is the formation of *blind valleys*, which are totally dry or contain streams which disappear underground at some point in the valley. Depending on the configuration and continuity of the underground drainage system, the subsurface water may reappear as springs downstream from the point of disappearance.

In humid warm or tropical regions karstification sometimes proceeds to a stage where soluble rocks are almost completely dissolved and carried away in solution. Under such circumstances the landforms take on an aspect quite different from those associated with intermediate karstification in which blind valleys, dolines, poljes, and uvalas predominate. Only scattered remnants of the soluble rocks are to be found in conical or domical hills or in mesa-like remnants. In contrast with karst dominated by open or closed depression landforms, the landforms stand above the surrounding land surface. Names applied to karst of this kind include *cockpit karst, tower karst,* or *hum* (hummocky) *karst,* depending on the sizes and shapes of the landforms residual from solution.

Several features of karst and the development of underground solution

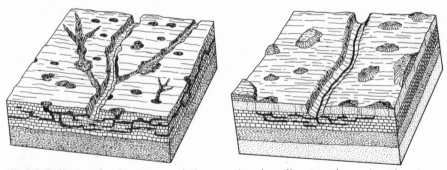

Fig.7-5. Dolines and subterranean solution openings in a limestone layer. A major stream crosses the area, but streams that at one time were tributary to the major stream disappear in the floors of uvalas.

Fig.7-6. Landforms (low hills and mesas) residual from extreme karstification in limestone in a humid tropical area. Subterranean solution openings have developed in a limestone exposed in the floor of a stream-cut valley by circulation of groundwater penetrating a jointed sandstone layer.

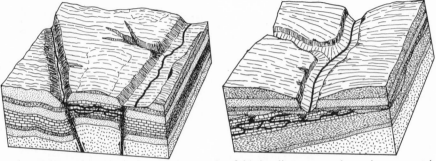

Fig.7-7. Valleys incised in fault zones intersecting folded sedimentary rocks resting on crystalline basement rocks. Underground circulation of water from the stream at the left to the stream at the right in a limestone layer has produced a complex system of interconnecting solution cavities (subjacent karst).

Fig.7-8. "Fossil" karst and solution cavities in a limestone layer below an angular unconformity. No indication of the existence of the limestone layer is given in bedrock exposures at the surface.

cavities are shown in Fig.7-5—7-8. The diagrams are highly idealized. In particular, underground caverns and other solutional features are shown as connecting within the planes of vertical sections, whereas, in actuality, the pattern of solution openings in three dimensions may be very complex. Underground solutional openings have almost an infinite variety of shapes, and shafts, galleries, and caves tend to be highly sinuous and anastomosing in both plan and cross-section.

Fig.7-5—7-8 are not intended to depict all of the kinds of features associated with karst and underground cave and shaft development in soluble rocks but, instead, suggest the wide range of conditions that lead to development of karst and subterranean cavities in soluble rocks.

REFERENCES

Herak, M. and Stringfield, V. (Editors), 1972. *Karst: Important Karst Regions of the Northern Hemisphere.* Elsevier, Amsterdam, 551pp.
Jennings, J. N., 1971. *Karst.* M.I.T. Press, Cambridge, Mass., 252pp.

Chapter 8

MECHANICS OF DAM FOUNDATIONS

INTRODUCTION

Application of the theory and practice of rock and soil mechanics in the design and construction of dams has reached an advanced state of sophistication. Unlike investigations for long underground tunnels, investigations of proposed dam and reservoir sites are concentrated within relatively small areas and, because of the threat to human life and property posed by failure of dams or their foundations, meticulously careful examination of foundation conditions and the geology of the reservoir site prior to design and construction is not only indicated but necessary.

Although every precaution may be taken to design and construct a dam with provisions for generous margins of safety within the dam itself during excavation and treatment of the foundation, and within the boundaries of the reservoir, it must be recognized that the dam and the reservoir behind it create dead-weight loads and water pressures that did not exist previously. Accordingly, the behavior of the dam, the materials in the foundations and the abutments and in the reservoir site require constant monitoring after construction and reservoir filling, so that short and long-term responses to the loads and water pressures can be ascertained. Completion of construction of a dam and reservoir does not terminate the responsibility of engineers and geologists for the safety of those who live and work in the drainage areas below dams.

Two outstanding methods, each with many variants, have been used in stress—strain analysis of dams and their foundations. In one method actual scale models of proposed structures are built and are subjected to stresses of a variety of kinds and intensities over short to long periods of time, and strains (displacements) are noted with strain instrumentation. The other method applies analytical mathematical theory and techniques and has been given great impetus in recent years by the development of the finite element method in structural and continuum mechanics (Zienkiewicz and Cheung, 1967; Zienkiewicz, 1968) which requires use of a computer in solution of problems.

Model studies and finite element studies require a thorough knowledge of geologic structure, petrography, and physical properties of the materials in foundations and abutments, as well as a knowledge of the forces (loading) exerted in the foundation by a dam and filled reservoir before a significant analysis can be made. Because of the requirements for quantitative data, increasing emphasis is being placed on foundation and abutment exploration by boreholes, geophysical measurements, and in situ rock mechanics tests as correlated with an increasingly refined and detailed location and interpretation of geologic features, especially rock discontinuities and anisotropism.

The major purpose in all design calculations and/or construction of models is limitation or control of displacements of rock masses under loading. An acceptable foundation or abutment is one in which it can be demonstrated that the available shear strength of a rock or soil mass exceeds by many times any shearing forces that may develop. A review of the causes of failure and/or serious deterioration in dams as documented in the literature indicates very clearly that they are almost always related to highly localized conditions and not to the mass average characteristics of large volumes of rocks or unconsolidated materials in foundations and abutments.

Computers do not have any built-in experience or capability for intuitive reasoning that will enable them to recognize serious flaws in foundations and will yield answers that are valid only in direct proportion to the quality and quantity of input. There usually remains, after completion of model or mathematical analysis of the foundation and abutments of a dam, considerable need for subjective appraisal of the relative significance of various geologic features based on a broad appreciation of the gross and more subtle factors that contribute to the behavior of earth materials under loads over short and long intervals of time.

STRESS AND STRAIN

Stress is the force per unit area acting to deform or cause a change in shape or volume of a body. *Strain* is the response to stress. Stress—strain relationships commonly are related to orthogonal x, y, and z reference axes or to polar coordinates. Many problems concerning stress and strain are solved in a plane and give solutions which are simpler and more comprehensible than solutions in three-dimensional space.

Stress acting perpendicular to a surface or an area is called *normal stress.* Stress parallel to a surface or an area is *shear stress.* Change in the length per unit of original length along a straight line is *longitudinai strain.* *Shear strain* is measured as the angular change in a right angle in a reference plane (xy, yz, or xz).

Materials may in general be classified as *elastic* or *non-elastic* with reference to their behavior under stress. Elastic materials, up to the point of failure, obey Hooke's law, that is, the strain is directly proportional to stress. Most natural materials that are classified as elastic in their behavior are not strictly so, but, if their behavior is *assumed* to be ideally elastic, many important stress—strain relationships can be examined and solved. There is a wide range of behaviors of materials under stress between so-called *perfectly elastic* and *perfectly plastic*

states. A perfectly plastic material will deform permanently without limit at some
critical or threshold stress and will not support a stress greater than the threshold
stress. In perfectly *viscous materials* (fluids) there is no threshold stress, and the
materials can not withstand any shear stress without deforming permanently.

Various kinds of idealized stress–strain relationships are shown in Fig.8-1,
and suggest the wide range of possible behaviors of natural materials under stress.

Strengths of materials commonly are tested in pressure apparatus in which
unconfined cylinders or cylinders confined by a fluid are subjected to uniaxial
stress to the point of failure. Tests employing a confining fluid commonly are called
triaxial tests, although the disposition of the applied stresses is not truly triaxial.
Examples of kinds of failure of cylinders under stress are shown in Fig. 8-2.
Fractures that develop in quasi-elastic materials at failure are generally of two
types: *lateral extension fractures,* which develop approximately parallel to the
direction of applied stress (Fig.8-2A) and *shear fractures,* which develop at from a
few to 45° to the direction of uniaxial stress in homogeneous materials (Fig.8-2B).
In perfectly isotropic materials tested in perfect cylinders the lateral extension

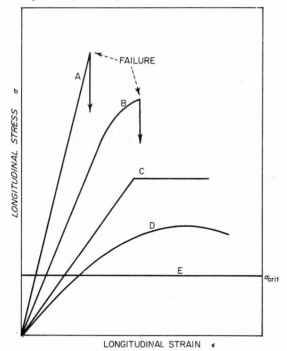

Fig.8-1. Idealized stress–strain curves. *A* = behavior of brittle elastic material which obeys
Hooke's law up to point of failure; *B* = brittle substance that is permanently strained just before
failure; *C* = ductile substance which demonstrates elastic response up to a critical stress and
permanent strain (plastic behavior) beyond critical stress; *D* = moderately ductile substance; *E* =
perfectly plastic material which will not deform if stress is less than critical value. Deforms
permanently without limit at critical stress.

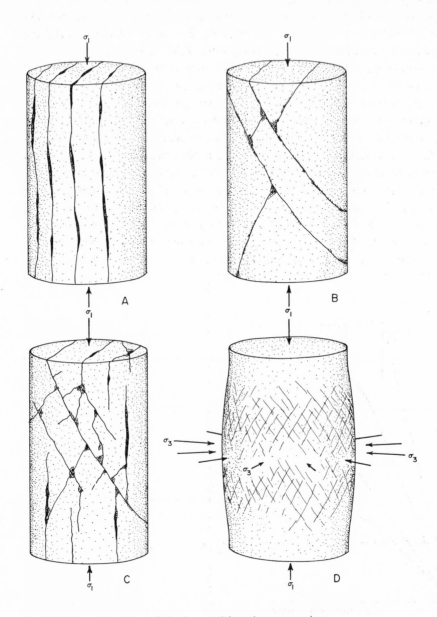

Fig.8-2. Failure of cylinders of elastic materials under compression.
A. Lateral extension fractures developed in a brittle substance under uniaxial stress.
B. Shear fractures developed in a quasi-elastic substance under uniaxial stress.
C. Failure with development of both shear surfaces and lateral extension fractures.
D. Failure by shearing of a cylinder under "triaxial stress".

fractures assume random azimuths (not as shown in Fig.8-2A), and the shear fractures form cones of rupture.

Fig.8-2C shows a cylinder which has failed by development of a combination of lateral extension fractures and shear fractures. Fig.8-2D illustrates development of shear fractures under "triaxial stress", that is, with a jacketed cylinder confined by fluid under pressure.

An example of lateral extension fractures in brittle sedimentary rocks is shown in Fig.8-3 and suggests the nature and origin of secondary permeability to

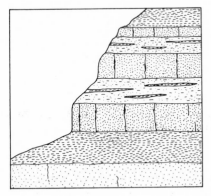

Fig.8-3. Lateral extension fractures in brittle sandstone layers between shale layers.

groundwater flow and the tendency for cliff-like exposures to develop in valleys in such rocks.

Most quasi-elastic natural materials do not exactly obey Hooke's law under stress, that is, stress—strain curves are not straight lines. Some materials show the phenomenon of *work-hardening* when subjected to alternately increasing and decreasing stresses below the limit of failure, such as the stresses that might be generated during repeated filling and drawdown of a reservoir. An idealized example of work-hardening is indicated in Fig.8-4. Initial application of stress produces a curved stress—strain curve, but relaxation of stress produces a curve that does not coincide with the original curve as it would if the stressed material were perfectly elastic. Subsequent increases and decreases in stress produce a family of curves that indicate a more nearly elastic behavior than existed when the material was originally stressed.

The behavior of materials such as the one depicted in Fig.8-4 has been explained as a consequence of irreversible closing or reduction of size of microfractures or micropores during initial application of stress. An additional possible factor is partial loss of pore water which by its presence tends to reduce ultimate strength.

Laboratory tests of cylinders under uniaxial compression usually are of short duration. However, it is well known that many elastic materials under constant stress over long periods of time behave more or less plastically up to a point of

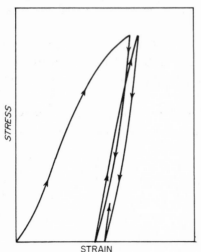

STRAIN

Fig.8-4. Stress—strain curves showing "work-hardening" of a material by repeated application and relaxation of stress. Stresses are below level that causes failure.

failure. This phenomenon is known as *creep* and has been observed especially in slabs of slate or marble subjected to stresses over periods of many years and in long-term laboratory experiments. An idealized strain—time curve is shown in Fig.8-5. A material under a constant differential stress just below the stress producing failure shows instantaneous elastic strain up to point *A*, decelerating *transient creep* from *A* to *B, steady creep* from *B* to *C*, accelerating *tertiary creep* from *C* to *D*, and finally, failure at *D*.

The particular effects that are observed depend on the material and the magnitude of the applied stress. In general the time interval for transient creep is at

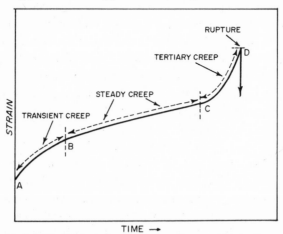

Fig.8-5. Idealized strain—time curve for a quasi-elastic material under constant differential stress. See text explanation.

most a few hours or few days. The time interval for steady creep is much longer and, for low stresses, may be of almost indefinite length. Accelerating tertiary creep is usually of short duration and occurs just before rupture. In very brittle substances the time lapse for tertiary creep is so short that it may not be observed.

EFFECTS OF CONFINING PRESSURE ON STRENGTH

The load of a dam on bedrock or unconsolidated foundation materials and the weight of water in a reservoir on underlying rocks or valley fill produce effects that depend on the magnitudes of the loads and the porosity and/or permeability of the substances subjected to the loads. In general, confining pressures produced by superjacent loads cause an increase in strength, whereas pore pressures associated with interstitial liquids tend to offset the effective confining pressure and thereby reduce the strength. Most bedrocks have primary permeabilities that are so low that pressures exerted by pore fluids are negligible. In contrast, unconsolidated deposits, especially those with interconnected pore spaces occupying more than about 5 vol.%, do not show notable increases in strength with confining pressures.

Similarly, unfractured bedrocks with low primary permeabilities increase in strength with confining pressure, more so than bedrocks intersected by interconnected, water-filled fractures or solution openings of any size.

The influence of confining pressure on strength has been studied intensively by means of short-time triaxial compression tests. Especially useful in interpreting the tests are the assumptions implicit in Mohr's concept of *internal friction* and the representation of stress–strain relationships in Mohr's circles of stress and strain (Handin, 1966; Obert and Duvall, 1967).

The shearing resistance τ_θ presented by an isotropic substance to failure is assumed to be the sum of a so-called *cohesive strength*, τ_0, and a term expressing frictional resistance to dislocation along a potential plane of failure. This term is the product of the *effective normal stress*, σ_θ, across the plane and the *coefficient of internal friction*, $n = \tan\phi$, in which ϕ is an angle analogous to the angle of ordinary sliding friction.

Thus:

$$\tau_\theta = \tau_0 + \tau_\theta \tan\phi \qquad\qquad (8\text{-}1)$$

The normal stress, σ_θ, and the shearing resistance, τ_θ (Fig.8-6), can be calculated by the equations:

$$\sigma_\theta = (\sigma_1 + \sigma_3)/2 - [(\sigma_1 - \sigma_3)/2]\cos 2\theta \qquad\qquad (8\text{-}2)$$

$$\tau_\theta = [(\sigma_1 - \sigma_3)/2]\sin 2\theta \qquad\qquad (8\text{-}3)$$

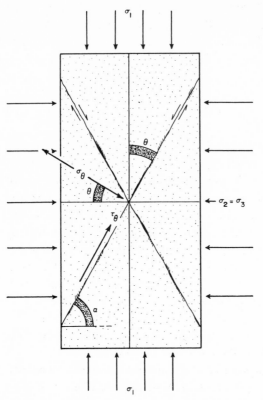

Fig.8-6. Longitudinal section of test cylinder under triaxial compression. Conjugate shears have developed at angle θ to direction of principal stress, σ_1.

Using eq. 8-1, 8-2, and 8-3 and solving for an angle θ (Fig.8-6) for which the shear stress is as high as possible when the normal pressure is a low as possible, we obtain:

$$\theta = \pm 45° \mp \phi/2 \tag{8-4}$$

The angle between symmetrically disposed planes of potential failure as seen in a section parallel to the axis of a test cylinder is bisected by the direction of maximum principal stress, σ_1. In triaxial compression tests a confining pressure is exerted equally in all directions normal to the walls of the test cylinder so that, in three dimensions, the surfaces of potential shear failure are cones with apical angles of 2θ. There are no directions of intermediate and least stress, and σ_2 equals σ_3.

Eq. 8-1–8-4 can be represented graphically by means of Mohr's stress circle for triaxial compression in which $\sigma_1 > \sigma_3$, as indicated in Fig.8-7 in which compression is assumed to have a negative sign. A circle is drawn with radius $(\sigma_1 - \sigma_3)/2$ and a center at $(\sigma_1 + \sigma_2)/2$ on a plot with orthogonal axes measuring normal stress and shear stress. The Mohr plot shows values of shear stress, τ_θ, and normal stress, σ_θ, acting on planes inclined at angles $\pm 2\theta$ to the direction of principal stress, σ_1, (eq.8-1 and 8-2). The angle of internal friction, ϕ, is formed by the tangent to the circle at P (Fig.8-7), and, n, the coefficient of internal friction can be obtained from $n = \tan\phi$ (or $n = \cot 2\theta$).

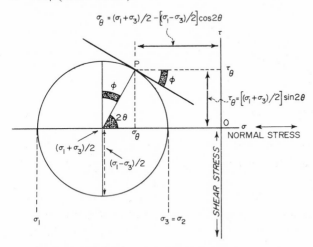

Fig.8-7. Mohr's circle for triaxial compressive tests.

In triaxial compressive tests it is common practice to make measurements at several confining pressures, and a succession of Mohr' stress circles are plotted on a single diagram, as in Fig.8-8. A line tangent to the several circles is called *Mohr's envelope* and fulfills the requirements of eq.8-1. Accordingly, the intersection of the envelope with the axis (Fig.8-8) yields a value for τ_0, the *cohesive strength*. Although the envelope is curved, indicating that $\tan\phi$ is not necessarily constant, in practice the envelope usually is assumed to be a straight line for ease in estimating the value of τ_0.

A particular value of Mohr's plots is their use in estimating by interpolation or extrapolation variations in the shearing resistance, τ_θ, as a function of any confining pressure. The assumption to be derived from Mohr's envelope (Fig.8-8) is that the envelope is a limiting curve, outside of which failure by shearing can be expected according to eq.8-1.

In nonisotropic materials the behavior under triaxial compression depends on the attitude of planes of weakness such as bedding planes in sedimentary rocks and foliation planes in metamorphic rocks with reference to the maximum principal stress direction. Cylinders in which the planes of weakness are at right angles to the direction of maximum principal stress rupture at higher stresses than cylinders in which the planes of weakness are parallel to the direction of maximum principal

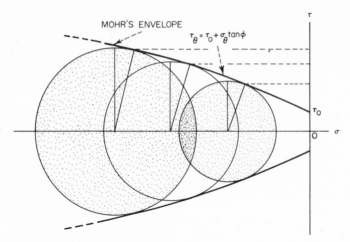

Fig.8.8. Mohr's circles and envelope obtained from successive triaxial compression tests with different confining pressures.

stress. In cylinders in which the planes of weakness make an angle with the maximum principal stress direction of about 15–60°, shearing tends to deflect into the planes of weakness in preference to cutting across them.

Handin (1969) points out that in cohesive materials internal friction is a fictitious quantity that can not be measured directly and emphasizes that it is not the friction associated with sliding on a newly developed fracture resulting from rupture under compression, nor is it the frictional resistance associated with slip on planar surfaces of any origin in rock bodies. Accordingly, application of the Mohr failure criterion in prediction of the response of fractured rock bodies to static or dynamic loading requires quantitative knowledge of the frictional properties of fracture systems as well as the mechanical properties of intact rock. Frequently, especially in complexly fractured rocks, significant critical data are unobtainable, either from field investigations or laboratory measurements, and prediction of behavior of rock masses under applied stresses must be based on subjective, qualitative estimates. Under such circumstances, field measurements of the actual response of rock masses to hydraulic jacking tests in tunnels or shafts provides a reasonable alternative to estimates based only on guesswork and conjecture.

Table 8-1 includes selected values for ultimate strength and fault angles (θ) for several rocks under short-term triaxial compression. The values are for room temperature and were taken mainly from a compilation by Handin (1966). The ultimate strengths reported are taken from the highest points on stress–strain curves for specified confining pressures and are the greatest stress differences that the materials being tested can withstand under the conditions of measurement. Examination of Table 8-1 clearly indicates an increase in strength with confining pressure.

The pressures in Table 8-1 are expressed as *bars*. A *bar* is a unit of pressure

TABLE 8-1

Selected data from short-time triaxial compression tests

Material	Confining pressure (σ_3) (bars)	Ultimate strength (bars)	Fault angle (degrees)
Basalt	0	2,620	21
	690	4,620	29
	1,030	5,510	21
Claystone	0	2,350	24
	690	4,320	27
	1,030	4,380	35
Dolomite	0	3,430	—
	490	4,410	—
	980	8,340	—
Granite	0	1,670	15
	490	4,710	20
	980	6,080	15
Limestone	0	530	30
	510	4,150	30
	1,020	3,130	27
Limestone	0	1,340	27
	230	2,010	29
	490	2,620	32
Quartzite	0	3,590	—
	1,010	10,790	25
	2,020	12,940	—
Sandstone	0	590	23
	350	1,610	36
	690	2,190	41
Sandstone	0	680	19
	280	2,000	35
	550	2,530	37
Shale	0	640	5
	260	980	—
	510	1,610	10
Siltstone	0	480	20
	500	1,470	30
	1,010	2,260	—

equal to 10^6 dynes/cm^2, also equal to the mean atmospheric pressure at about 100 m above mean sea level. Bars can be converted to psi by multiplying by 14.7 or, for approximate conversion, by 15.

PRESSURES ASSOCIATED WITH DAMS AND RESERVOIRS

Construction of a dam and filling of the reservoir behind it create load stresses on the floor and sides of a valley that did not exist previously. An analysis of these stresses is an urgent prerequisite to dam construction and reservoir filling, so that there is ample assurance that there will be no possibility of structural failure because of foundation conditions.

The kinds and distributions of imposed stresses created by a dam on its foundation depend on the shape of the dam and the materials used in its construction. Dams built of masonry or concrete can be considered to behave as cohesive, rigid, monolithic structures, although minor adjustments may occur by quasi-elastic dislocations and by movements along construction joints. The stress acting on the foundation is a function of the gross weight of the dam as distributed over the total area of the foundation on which the dam rests. In contrast, earth and rock-fill dams exhibit gross semiplastic behavior, and the pressure on the foundation at any point depends on the thickness of the dam above the point.

The pressures exerted by earth and rock-fill dams resemble in some respects those exerted by the water in a reservoir, but pressure distribution is modified by the fact that the materials of construction have some inherent strength, and fail only after some threshold stress has been exceeded. Pressures exerted by water in the reservoir behind a dam are hydrostatic and increase linearly with depth.

Pressure is a scalar quantity and is expressed in various units. The *density* of a substance, ρ, is its mass divided by its volume and is a ratio. Pressure often is expressed in terms of a force exerted by a unit cube of a substance in the gravitative field. In engineering practice commonly used units of pressure are psi (lbs/inch^2), psf (lbs/ft^2), g/cm^2 (grams/square centimeter), and kg/m^2 (kilograms per square meter). Numbers that are frequently used are the weight of a cubic foot of water (approximately 62.5 lbs) and the psi of water as a function of depth (0.433 psi per added foot of depth). The weight of a cubic foot of concrete is approximately 162.5 lbs.

The difference between hydrostatic pressures and pressures exerted by an essentially rigid structure such as a concrete dam is shown in Fig.8-9. In Fig.8-9A pressures are hydrostatic and increase with depth. On the assumption that the pressures are directed normal to the floor and sides, they are shown as vectors of increasing magnitude with depth. Of course, fluctuations in the water level are attended by changes in pressure at any particular point below water level in the reservoir, thus creating a dynamic system of changing stresses. In Fig.8-9B the dead-weight load of a concrete dam is distributed over the total area of the foundation and is shown by vectors normal to the surface beneath the dam. The situation depicted in Fig.8-9 is essentially a static one, and depends only on the weight of the dam and the area of the foundation.

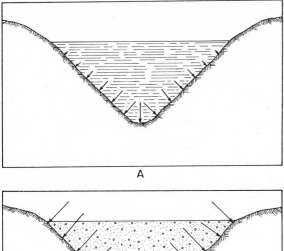

Fig.8-9. Pressures (stresses) on a foundation in a valley.
A. Pressures due to water in a reservoir.
B. Pressures from the weight of a rigid concrete dam.

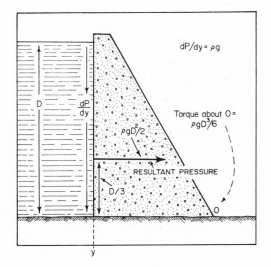

Fig.8-10. Forces acting on a rigid dam owing to hydrostatic pressures.

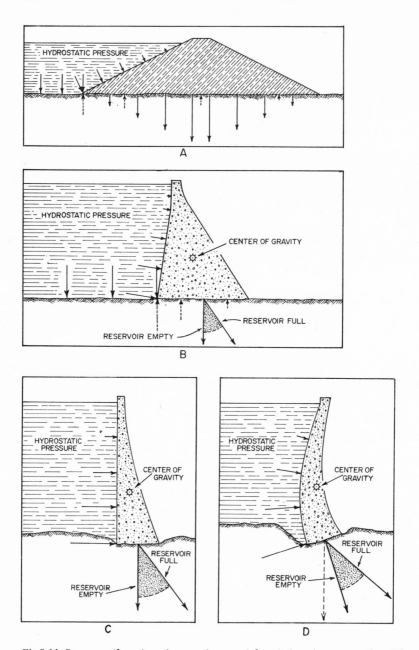

Fig.8-11. Pressures (forces) acting on dams and foundations in cross-section. Where shown, dashed arrows indicate uplift pressures associated with seepage beneath the dams.

A. Earth dam.

B. Concrete dam.

C. Arch dam with vertical upstream face.

D. Arch dam with curved faces both in plan and cross-section.

Water exerts hydrostatic pressures not only on the floor and walls of a reservoir but also on the upstream face of a dam. An analysis of these pressures is made with the aid of Fig.8-10, in which D is the depth of water in a reservoir, and P is hydrostatic pressure per unit area acting on the vertical face of a concrete dam assumed to behave as a rigid body. The change in pressure with depth (in the y direction) is given by:

$$dP/dy = \rho g \qquad\qquad\qquad\qquad (8\text{-}5)$$

in which ρ is the density of water (changes slightly with temperature), and g is the acceleration due to gravity. In calculations ρg can be set equal to the pressure exerted on a unit surface by a unit cube of water. By integration, the resultant pressure exerted on a unit thickness of the dam proves to be $\rho g D^2/2$, the resultant torque about O is $\rho g D^3/6$, and the line of action of the resultant force acting on the dam is $D/3$ above the base of the dam. If the width W of the body of water behind the dam is known, the total force and torque can be calculated by simple multiplication.

In calculations of the stability of the dam the torque tending to rotate the dam about O should be added to the tendency of the dam to be rotated in the same direction about the same point by uplift forces related to seepage beneath the dam (see Chapter 5).

Several cross-sections are shown in Fig.8-11 and indicate the kinds of pressures and the nature of their interaction in dams. The cross-sections are highly diagrammatic and do not take into account three-dimensional variations in foundation pressures which depend on the configuration of the dam and the fact that the foundation of the dam slopes upward toward the abutments, thus complicating two-dimensional stress analysis.

Fig.8-11A illustrates an earth dam, a nonrigid structure that under stress behaves semiplastically. Because of relatively easy internal adjustments to loads, the pressures exerted on the foundation are approximately equal to the weight of overlying prisms of material of different heights. Pressures exerted on the dam by water in the reservoir tend to cause greater adjustments near the base of the dam than at shallower depths.

Fig.8-11B is a cross-section of a concrete gravity dam, presumed to behave as a rigid body. When the reservoir is empty, the weight of the dam is directed vertically downward. When the reservoir is full, a combination of hydrostatic pressure on the upstream face of the dam and the weight of the dam produces a force vector inclined downstream away from the vertical force vector, and there is a tendency for the dam not only to be displaced downstream but also to rotate about the downstream toe of the dam because of a torque.

Fig.8-11C and D show force vectors for empty and filled reservoirs behind concrete arch dams. Again, the representation is highly diagrammatic. Unlike gravity dams, arch dams, because of the egg-shell effect, tend to resist downstream dislocation, and the displacing forces, instead, are transmitted laterally, at least in part, through the dam and toward the abutments.

MECHANISMS OF FOUNDATION FAILURE

In general, failures in flat, nearly horizontal foundations of earth and rock-fill dams are not the result of shearing dislocations owing to the load of the dam. Instead, foundations give way because of inadequate treatment for seepage, either within the dam or beneath it, or as a consequence of construction of a dam on a foundation which slopes steeply upstream or downstream.

Dams built of concrete require careful investigations of foundations because of the concentrated nature of loads exerted by them in their behavior as essentially rigid structures. There is a vast literature on the interaction of loads produced by structures and soil or rock foundation, on which they are built, and no effort is made here to review the many theories and empirical relationships that have become a part of the modern science of soil mechanics. Instead, attention will be given to some simple concepts that have been successfully applied and to geological conditions in foundations that are unfavorable in the light of these concepts.

Fig.8-12 shows a mechanism of foundation failure under load that has been widely employed in the analysis of bearing strengths of soils (Jumikis, 1962; Leonards, 1962; Talobre, 1967). This mechanism responds to analysis by application of Mohr's theory of shear failure described earlier in this chapter, and is sometimes called the "plastic method of analysis of bearing capacity".

For the mechanism to work, a portion of the foundation must behave plastically (Fig.8-12). However, this presents no problem because most natural materials, including so-called elastic rocks, usually show some plastic behavior in masses because of the presence of micro or mega fractures or pore spaces.

For a load directed vertically downward as in Fig.8-12A it is assumed and born out by model studies that a symmetrical wedge is formed by shear dislocation. Extension of the shear surfaces bounding the wedge, first as curved surfaces, and then as planar surfaces intersecting the earth's surface, provides a mechanism for shear failure along symmetrically disposed surfaces beneath the dam.

Fig.8-12B shows the consequences of application of an inclined load, either because of an inclined surface at the base of the dam or because of an interaction between the load component acting vertically due to the weight of the dam and the pressure exerted on the dam by the hydrostatic pressure of the water in the reservoir. The result is a tendency to shear along a single surface that intersects the earth's surface downstream from the dam. If pre-existing planes of weakness of-

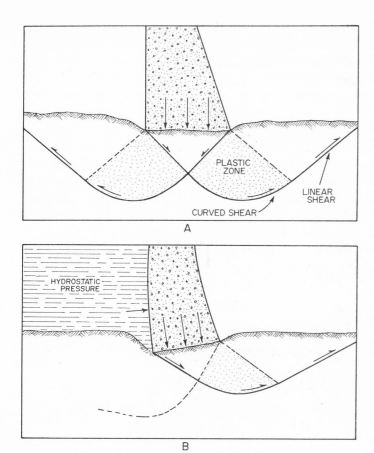

Fig.8-12. "Plastic" mechanism for shear failure of foundation materials.
A. Load is directed vertically.
B. Load is directed asymmetrically.

appropriate orientation exist in the foundation materials, it should be expected that the shear surfaces will deflect into them.

GEOLOGIC CONDITIONS PROMOTING FOUNDATION FAILURE

Geologic conditions in foundations for concrete dams that should be avoided are indicated in Fig.8-13. The hazards that are associated with these conditions should be related to the diagrams in Fig.8-10—8-12. The geometry of the planes of weakness in each of the depicted situations is such that failure according to the mechanism implied in Fig.8-12 is possible, indeed likely. The conditions shown in Fig.8-13D are similar to those that caused the disastrous failure of the Malpasset dam in France in 1959 (Talobre, 1967).

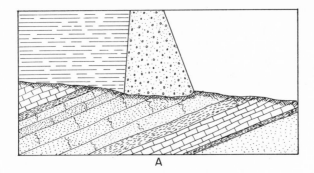

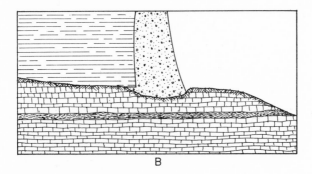

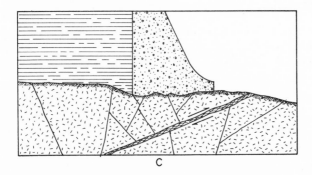

Fig.8-13. Geologic conditions promoting failure of foundations of concrete dams.
A. Brittle, fractured sandstones rest on a weak shale layer dipping upstream.
B. Horizontally layered limestones rest on a weak shale layer which extends downstream to a steep slope in the valley floor.
C. Fractured crystalline rocks lie above a flat fault containing sheared, gougy materials of very low strength.

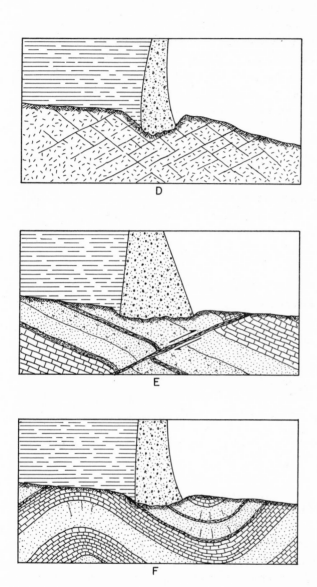

D. Intersecting strong conjugate joints have attitudes that promote easy mass shear dislocations.
E. Sedimentary rocks dipping downstream are intersected by a fault dipping upstream and containing materials of low strength.
F. Folded rocks containing thin, weak layers of shale present a potential for foundation failure.

Slope failures toward abutments (in direction of the dam axis) which disturb or dislocate the abutments are rare. In concrete dams in which slopes in the abutment areas maintain themselves during excavation for the foundation, the possibility of downslope movement along surfaces that intersect the foundation of the dam is remote because of the added stability provided by the weight and strength of the dam. However, the possibility exists that slopes *above* the dam, especially in deep valleys, may fail and bury surface structures with rock and/or soil debris.

Slopes beneath and above earth and rock-fill dams, if sufficiently steep and weakened by infiltration of groundwater, may fail along surfaces that intersect the abutment portions of dams. An example portraying this situation is indicated in Fig.8-14 in which a sandstone layer containing vertical fractures may provide sufficient load to cause development of curved shear surfaces in an underlying water-soaked shale layer.

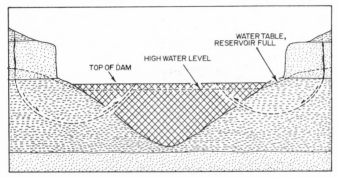

Fig.8-14. Conditions promoting possible slope failure beneath abutments of an earth or rock-fill dam along curves shear surfaces. Shale beneath a sandstone layer has been weakened by infiltration of water from reservoir.

TESTING OF STRENGTH OF FOUNDATION MATERIALS

An essential prerequisite to design and construction of a dam is determination of the strength of foundation materials, especially if there is some doubt as to their ability to support without dislocation the loads superimposed on the foundation by the dam and by the water in the reservoir behind the dam. It should be expected that strengths of masses of rock as determined by in situ measurements generally are less than strengths determined in the laboratory on test cylinders because of the common presence of fractures and other planes of weakness in large volumes of foundation materials. An additional complicating factor is anisotropism in most natural materials which exerts directional controls in strength that may be difficult to assess. For example, strengths measured by stresses exerted normal to layering in sedimentary or metamorphic rocks in cylinders in the laboratory usually prove to be much greater than strengths in test cylinders which are parallel to these features.

In design of engineering structures common use is made of the *modulus of elasticity, E,* called Young's modulus, *Poisson's ratio, ν,* and unit weight or density, *ρ.* All of these parameters can be measured in test cylinders in the laboratory or by a variety of methods in the field. Stress is signified by *σ* and strain by *ε*.

The modulus of elasticity usually is defined as the ratio of longitudinal stress to longitudinal strain, that is, as σ/ϵ, and is a measure of the ability of a material to resist directed stress. In conventional tests in pressure machines the modulus of elasticity is determined by measuring the shortening of a test cylinder under a given applied stress at the ends of the cylinder.

When a cylinder (or bar) is compressed, shortening is accompanied by lateral extension in directions at right angles to the direction of the applied stress. Measurement of the lateral extension enables calculation of Poisson's ratio. Suppose that the direction of applied stress is parallel to the *x*-axis and lateral extension is measured parallel to the *y*-axis of rectangular coordinates. Then Poisson's ratio is expressed as $\nu = \epsilon_y E/\sigma_x$.

TABLE 8-2

Ranges of moduli of elasticity of selected rocks[1]

Rock types	Elasticity	
	(psi × 10^6)	(bars × 10^5)
Igneous rocks		
Granite	1.5–11.9	1.0– 8.2
Diorite	3.6– 6.1	2.5– 4.2
Andesite	4.7– 6.9	3.3– 4.8
Basalt	5.9–12.4	4.1– 8.5
Diabase	10.2–13.9	7.0– 9.6
Metamorphic rocks		
Quartzite	1.2– 6.4	0.8– 4.4
Greenstone	3.4–15.2	2.3–10.5
Gneiss	3.5–15.1	2.4–10.4
Amphibolite	6.7–15.1	4.6–10.4
Sedimentary rocks		
Gypsum	0.17– 1.1	0.12– 0.8
Shale	0.3– 9.9	0.2– 6.8
Limestone	0.4–14.1	0.3– 9.7
Marlstone	0.4– 7.0	0.3– 4.8
Sandstone	0.6– 8.0	0.4– 8.0
Siltstone	1.0– 9.3	0.7– 6.4

[1] Conversion factors: 1 bar = 14.5 psi = 1.02×10^4 kg/cm².

The modulus of elasticity and Poisson's ratio can be measured in the laboratory or in the field by a variety of methods. Measurement of these parameters with pressure machines characterizes the so-called "static method". Increasing use is being made of methods in which measurements are made of the velocities of transmission of artificially induced shock waves, the so-called "dynamic methods".

Some representative values of the modulus of elasticity, taken from a variety of sources, are shown in Table 8-2. When the modulus of elasticity and the density are known, Poisson's ratio can be calculated. In fact, when any two of the three parameters are known or measured, the third can always be calculated.

DYNAMIC METHODS FOR LABORATORY MEASUREMENT OF ELASTIC CONSTANTS

Outstanding methods for measuring the modulus of elasticity by dynamic methods in the laboratory utilize well-known relationships among the velocity of transmission of artificially induced shock waves (seismic impulses), the modulus of elasticity, E, Poisson's ratio, v, and the density, ρ. The velocity of transmission differs with the kinds of waves that are being transmitted, and, in anisotropic substances, with the direction of transmission. Two kinds of waves usually are employed in measurements: *compressional waves* and *shear waves*.

Following are basic equations relating velocities of seismic impulses of compressional and shear waves to elastic constants:

$$V_p = [\{E(1-v)\}/\rho(1+v)(1-2v)]^{\frac{1}{2}} \tag{8-6}$$

$$V_s = [E/2\rho(1+v)]^{\frac{1}{2}} \tag{8-7}$$

in which V_p = velocity of the compressional wave, V_s = velocity of the shear wave, and $V_p > V_s$

Expressions for the elastic constant, E, and Poisson's ratio, v, involving only velocities, V_p and V_s, and the density, ρ, are as follows:

$$E = V_s^2 \rho [\{3(V_p/V_s)^2 - 4\}/\{(V_p/V_s)^2 - 1\}] \tag{8-8}$$

$$v = \tfrac{1}{2}[\{(V_p/V_s)^2 - 2\}/\{(V_p/V_s)^2 - 1\}] \tag{8-9}$$

Two dynamic methods of evaluating the elastic constants by the dynamic method in the laboratory are in widespread use. In one method instantaneous seismic impulses are generated by a piezoelectric crystal and the velocities of transmission of compressional and shear waves along a cylinder or bar are measured. In the other method a transducer is utilized to cause vibrations in test samples. By a tuning mechanism a critical frequency of transmission is identified and enables calculation of the elastic constants. Designs of equipment and procedures for dynamic measurement of elastic constants, together with formulas for correction of measurements, are given in detail by Obert and Duvall (1967).

FIELD TESTING OF FOUNDATION ROCKS

Although laboratory measurements of the physical properties and elastic constants of foundation materials yield data of considerable utility in design of dams, the fact that should be kept uppermost in mind is that dams rest on foundations of rocks or unconsolidated materials and not on laboratory test cylinders. Primary anisotropism owing to the origin of the material, and secondary anisotropism coming about from development of fractures and folds, and from alteration by solutions of shallow or deep-seated origin induce weaknesses that are not usually accounted for in indoor laboratory tests. Experience has demonstrated that bulk strengths of large masses of rock, almost without exception, are less than strengths of samples subjected to laboratory tests by a small to large factor.

Existing stresses in foundation rocks prior to dam construction and reservoir filling are of two kinds (1) stresses associated with the dead weight of materials in slopes and the valley floor as modified by the slope configuration and internal structures such as fracturing and layering, and (2) residual or tectonic stresses. Residual stresses are unbalanced stresses that are preserved in elastic or quasi-elastic rock bodies and were generated at some time in the geologic past when the rock bodies were subjected to deformation by earth forces. Tectonic stresses are unbalanced stresses that are related to elastic strain of rocks by presently active earth forces, such as those that build up in intensity along faults in regions of modern-day earthquake activity. Many residual stresses originally were generated by tectonic dislocations of the earth's crust, but relaxation of the regional forces active at the time of the dislocations was not accompanied by total relaxation of stress within the rock bodies subjected to the forces.

Residual and tectonic stresses (and associated strains) are superimposed on and interact with dead-load stresses (and strains) dependent primarily on the force of gravity and, in some instances, cause unexpected behavior of rocks during foundation excavation or when the foundation and abutments are loaded with a dam and the water in the reservoir behind it. Well-known examples of release of tectonic stresses, whether residual or not, have been noted in quarrying operations containing "popping" rock.

It is clear that in situ measurements of elastic properties of elastic or quasi-elastic foundation and abutment rocks should take into account the possible existence of residual or tectonic stresses.

Of particular concern in construction of dams and reservoirs is the mass or bulk strength of foundations and reservoir slopes. The strength of a foundation is measured or estimated to determine its bearing strength or capacity under the weight of a dam and the hydrostatic pressure of stored reservoir water, and to assess the interaction of the foundation with stresses that fluctuate with filling and drawdown of the reservoir. Knowledge concerning the foundation strength may eliminate consideration of construction of a dam, or, if a dam is built, will enable calculation of an adequate factor of safety.

Especially useful in estimating bulk strength and behavior under loads of foundation and abutment rocks are data obtained from vertical and inclined boreholes, shafts, and tunnels into the abutment. Equipment is available which measures elastic constants of rocks in boreholes, and rupture strengths can be determined by hydraulic fracturing by fluids introduced at high pressures between firmly set packers in boreholes. Some knowledge of the elastic constants and the state of stress also is obtained from measurement of velocities of induced seismic waves through rock masses.

Probably, the most effective and revealing method of estimating strength of foundations utilizes shafts and tunnels driven into rock in the foundation and into the abutments. Not only do these excavations permit measurement of bulk elastic constants and rock strength by rock-mechanics instrumentation, but they permit detailed visual observation of rock characteristics, including lithology, fracture patterns, and the effects of weathering or other kinds of rock alteration.

A prime objective of investigations in shafts and tunnels is estimation of the response of foundation materials to loading at the earth's surface. In earth and rock-fill dams minor adjustments in foundations, as the loads of dams and reservoirs are built up, generally are expected and of slight consequence. However, settlement of foundations or lateral and vertical dislocation of abutment rocks by concrete structures is of utmost concern because of the possibility of dam failure. Uneven nonelastic adjustments in foundations in general pose more of a threat to the integrity of a dam than systematically distributed elastic dislocations.

Measurements of bulk elastic constants and strengths in a foundation and location of test holes or excavations should be carefully correlated with the geology of the foundation in plan and cross-sections. Particular attention should be given to locations and attitudes of planes of weakness and locations of bodies of rock with low original strengths.

In a typical investigation of a dam foundation, particularly in a rock foundation where a high bearing capacity is required for safe construction of a dam, field and laboratory measurements of rock properties and detailed interpretation of bedrock geology leave some questions unanswered, at least in a precise quantitative sense. Accordingly, in design and construction, particular care should be exercised in remedial treatment of the foundation and design of the dam to provide a generous factor of safety.

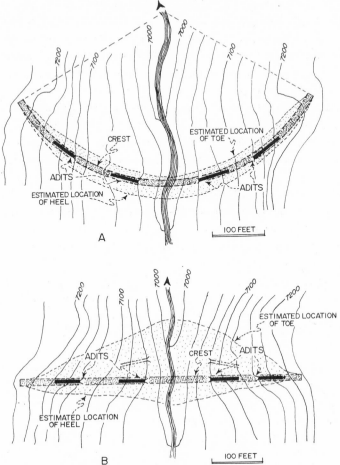

Fig.8-15. Suggested locations for exploratory adits (tunnels) in dam foundation for inspection of geological features and measurement of mass properties by rock mechanics instrumentation.
A. Concrete arch dam.
B. Concrete gravity dam. Possible alternate locations for adits in lower slopes are indicated by dashed lines.

Some possible locations for exploratory adits (tunnels) in foundations of concrete dams are shown in Fig.8-15. The adits should be long enough to penetrate rock well below the zone of surface fracturing and weathering, with the expectation that the weathered and fractured materials will be removed during excavation for the keyway for the dam. The number of adits is determined by the height of the dam, its design, and the degree of uncertainty as to the detailed geology and mass physical characteristics of the foundation rocks.

A great variety of methods are available for field stress—strain measurements, including measurements of residual and tectonic stresses (Obert and Duvall, 1967; Talobre, 1967; Stagg and Zienkiewicz, 1968; Jaeger and Cook, 1969).

SEISMIC METHODS FOR MEASUREMENT OF ELASTIC CONSTANTS IN THE FIELD

Eq.8-6–8-9 provide a basis for field measurements of elastic constants by seismic methods as well as measurement of these constants in the laboratory in test cylinders or bars. For use in boreholes equipment has been devised which emits gamma rays and enables measurement and logging of variations in density of formations in the walls of the core hole. Another kind of equipment, activated by a transducer, measures and logs velocities of compressional and shear waves in the materials intersected by the drill hole, and, taken together with the density data, enables preparation of a log showing variations with depth of the modulus of elasticity and Poisson's ratio.

In field seismic investigations at the surface by the refraction method only the velocity of compressional waves ordinarily is recorded. However, if average values of the density and Poisson's ratio are estimated or are measured by the borehole method described above, it becomes possible by eq.8-6 to calculate the modulus of elasticity for the rock mass traversed by the seismic impulses. This method has a particular value in that the modulus of elasticity so determined is the modulus for a mass of rock rather than the modulus for a narrow interval of a borehole or the modulus of a cylinder or bar tested in the laboratory.

REFERENCES

Handin, J., 1966. Strength and ductility. In: S. P. Clark Jr., (Editor), *Handbook of Physical Constants. Geol. Soc. Am., Mem.,* 97: 223–289.
Handin, J., 1969. On the Coulomb–Mohr failure criterion. *J. Geophys. Res.* 74: 5343–5348.
Jaeger, J. C. and Cook, N. G. W., 1969. *Fundamentals of Rock Mechanics.* Methuen, London, 513pp.
Jumikis, A. R., 1962. *Soil Mechanics.* Van Nostrand, New York, N.Y., 791pp.
Leonards, G. A. (Editor), 1962. *Foundation Engineering.* McGraw-Hill, New York, N.Y., 1136pp.
Obert, L. and Duvall, W. I., 1967. *Rock Mechanics and the Design of Structures in Rock.* Wiley, New York, N.Y., 650pp.
Stagg, K. G. and Zienkiewicz, O. C., 1968. *Rock Mechanics in Engineering Practice.* Wiley, New York, N.Y., 442pp.
Talobre, J. A., 1967. *La Mécanique des Roches.* Dunod, Paris, 442pp.
Zienkiewicz, O. C. and Cheung, Y. K., 1967. *The Finite Element Method in Structural and Continuum Mechanics.* McGraw-Hill, London, 272pp.

Chapter 9

GEOLOGICAL AND GEOPHYSICAL INVESTIGATIONS OF DAM AND RESERVOIR SITES

INTRODUCTION

The contents of this chapter presuppose that the reader is familiar with text-materials of foregoing chapters. Additional material not present in preceding chapters includes descriptions and discussions of the theory and instruments employed in modern geotechnical study and evaluation of dam and reservoir sites prior to, during, and after construction. The point-of-view is that of a geologist charged with the responsibility of supplying engineers, including rock mechanics experts, with geological and geophysical data required for safe design and construction of a dam and preservation of the integrity of the reservoir behind it.

GEOLOGICAL AND GEOPHYSICAL INVESTIGATIONS AND MATERIALS INVENTORY

Examination of a possible site for construction of a dam and reservoir requires investigations of a variety of kinds. Initially, efforts are coordinated to determine whether construction of a dam and appurtenant features is economically feasible by giving consideration to the geology of the dam and reservoir site, the kinds and locations of materials available for construction of the dam, and economic, hydrologic, human, and geographic factors.

The geologic report prepared at this stage of investigation should include a map showing all geologic features visible at the surface at the dam site, within the reservoir, and in immediately contiguous areas. The map should be an accurate representation on a topographic base of what can actually be seen in the field and indicates in plan the kinds and distribution of materials such as bedrock exposures, unconsolidated deposits, and vegetation and water cover. Extrapolation or interpolation beneath surface cover of features observed in bedrock exposures such as bedding, joints, and faults should not be attempted except where the existence and location of these features beneath the surface cover is unquestionable

Construction of accurate geologic cross-sections across and parallel to the axis of the dam generally requires exploratory drilling and geophysical measurements. Where it is presumed that the dam will rest on unconsolidated deposits drive-samples are obtained for testing, and water tests are made to determine permeability. In situations where all or part of the dam will be constructed on bedrock, the foundation is explored by core drilling and water testing. Ordinarily, recovery of drive-samples and rock cores is preceded by geophysical investigations, the results of which are correlated with drill-hole data in constructing geologic cross-sections. However, under appropriate conditions, valuable information may be derived from geophysical logging of the boreholes.

Feasibility-stage investigations are terminated when it is concluded that a dam and reservoir site is or is not acceptable as a location for construction. Assuming that a site is acceptable, the emphasis is now directed toward obtaining data that will be required for adequate design and close estimation of quantities and construction costs, including a detailed assessment of the quality and quantity of available construction materials.

Various geologic and geophysical techniques of investigation and geologic factors that should be given close attention are outlined in subsequent sections of this chapter.

REMOTE SENSING IMAGERY IN GEOTECHNICAL INVESTIGATIONS

Images obtained from remote sensing include aerial photographs, photographs from orbiting space craft, radar images, and infra-red images. Photographs are filtered or unfiltered standard black-and-white, standard color, filtered color, and colored infra-red, each serving a particular purpose in definition of features at the earth's surface. Radar imagery is obtained by translating reflected impulses generated by a scanning radar antenna into a photographic image showing relief features of the earth's surface. Infra-red thermal imagery detects slight differences in heat emission and absorption at the earth's surface and in bodies of water. Commonly thermal images are obtained just before dawn and slightly after sunrise.

Images from remote sensing play an important but subsidiary role in the interpretation of surface geology and should be considered only as tools assisting in the mapping of geology by careful ground observation. Preparation of geologic maps and sections from imagery alone, especially by persons who have limited field experience, should be avoided except for areas where surficial landforms and bedrock structures are of the simplest and most obvious kinds. Contrary to the convictions held by some, there is no acceptable substitute for examination and mapping of topographic and geologic features on the ground by the time-consuming and sometimes tedious methods of conventional field geology. Overly-zealous efforts in interpretation of details of geology from imagery alone frequently result in maps that require a greater effort in substantiating or disproving their validity than would have been expended by original mapping in the field using imagery merely as a basis for field location of features of geologic interest.

It is abundantly evident that planning of a program of subsurface exploration by geophysical techniques and boreholes requires an accurate geologic map that is based entirely on direct, on-the-ground field observation and not on conjecture or ill-founded interpretation of images obtained from remote sensing.

GRAPHICAL AIDS

Many problems associated with the space geometry of lines and planes can be characterized and solved by the use of the techniques of descriptive geometry. Of particular use in geologic investigations are stereographic and equal-area projections which enable representation by points on a flat surface of the bearings and inclinations of planes or lines such as bedding planes, foliation planes, faults, joints, and intersections of planes.

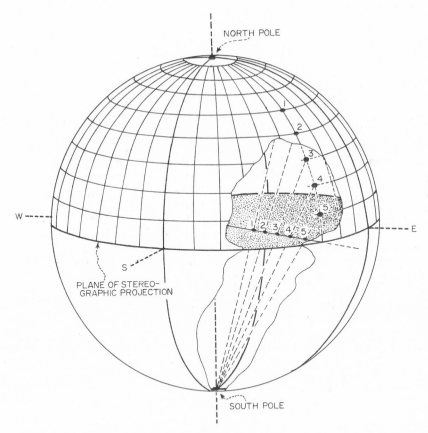

Fig.9-1. Stereographic projection of points on a spherical projection into the horizontal diametral plane (stippled).

Fig.9-1 indicates the basis for construction of a stereographic projection. Points *1, 2, 3*, etc., on a surface of a sphere have identifiable latitudes and longitudes (azimuths). Projection of these points into a horizontal diametral plane (stippled) by extending a line from each point to the south pole of the stereographic projection yields a stereographic projection of the points.

Stereographic projection of lines of latitude and longitude (azimuth) into a horizontal or vertical diametral plane of the spherical projection provides a *stereographic net* such as the one illustrated in Fig.9-2. All lines on the net are circles, arcs of circles, or straight lines. The intersection of a normal to a plane or any line passing through the center of the spherical projection with the surface of the sphere defines the *pole* of the plane or line. The pole for a plane or line of known angular relationships can also be plotted on the stereographic net, most conveniently on a rotatable vellum overlay on the net. If required, poles may be rotated to new positions over the surface of the net so as to solve many kinds of problems encountered in geological studies.

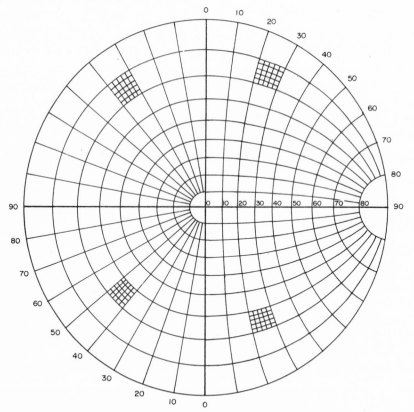

Fig.9-2. Stereographic net for plotting data in stereographic projection. Left side, polar net; right side, equatorial net.

The equal-area net has a particular use in the representation and analysis of large numbers of measurements of strikes and dips of planar features. The net (Fig.9-3) superficially resembles the stereographic net, but is constructed in a different manner so that at a given latitude similarly enclosed areas are the same or nearly the same.

The difference in the methods of locating the pole of a plane in a stereographic net and in an equal-area net is indicated in Fig.9-4. P is the pole of the plane on the spherical projection in a vertical section with an azimuth (longitude) that includes the normal to the plane. P' is the projection of P in an equal-area net such that the distance from the center of the net, OP' is equal to $PN/1.414$. In stereographic projection the distance OP'' is equal to $R \tan\rho/2$, in which R is the radius of the sphere and ρ is the angle of dip of the plane.

A common use of the equal-area net is suggested by Fig.9-5. Poles of a large number of joints with strikes and dips measured in the field have been plotted on a vellum overlay on an equal-area net. A distinction is made on the plot between joints with rough surfaces and joints with smooth surfaces.

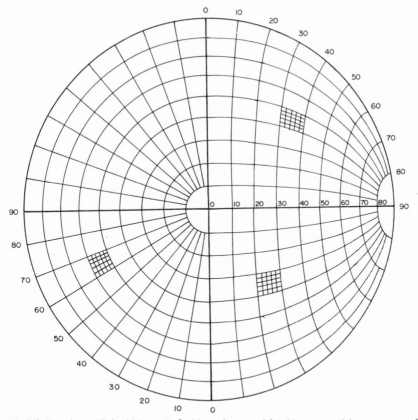

Fig.9-3. Equal-area (Schmidt) net. Left side, polar net; right side, equatorial net.

   To statistically analyze the distribution of the poles of the joints, use is made
of a point-counter with a circular opening the radius of which is equal to 0.1 of the
radius of the net (inset, Fig.9-5). The area of the circular opening in the counter is
one per cent of the total area of the net, and, accordingly, the counter is called a
"one per cent counter". The counter is moved over the plot, and the numbers of
poles within the circular opening is noted on the vellum overlay and converted to
percentages of the total number of poles at a sufficient number of locations to
enable drawing of contour lines of equal percentage concentration of poles. On the
periphery of the net poles are counted by placing the center of the counter at
opposite ends of diameters of the net and totalling the poles. In Fig.9-5 only the
10% contour is shown.

   Great circles are drawn $90°$ from the centers of gravity of the areas
enclosed by the 10% contours in Fig.9-5 and indicate the approximate strikes and
dips of two sets of joints corresponding to the concentrations of poles enclosed by

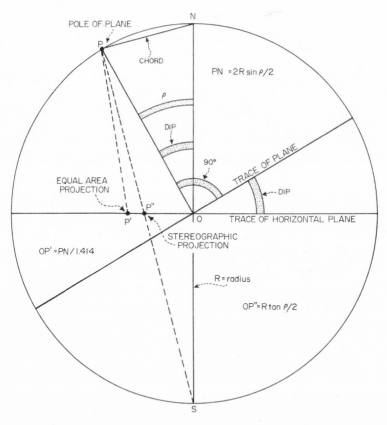

Fig.9-4. Comparison of methods for locating poles of a plane in equal-area projection and in
stereographic projection.

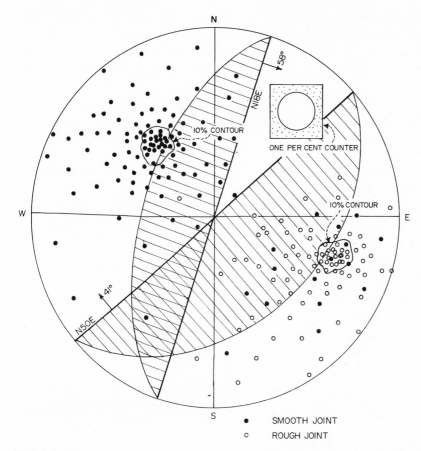

Fig.9-5. Equal-area plot of poles of joints. Hatched plane indicates approximate strike and dip of joints with poles enclosed by the 10% contour. See text discussion.

the 10% contours. Physically the diagram can be interpreted to mean that at least 10% of the total number of joints observed in the field have strikes near N18E and dips near 58° to the southeast and that at least 10% of the joints have strikes near N50E and dips near 41° to the northwest. Further, the northwest-dipping joints are dominantly smooth, and the southeast-dipping joints are dominantly rough.

BOREHOLE OPERATIONS

Thorough investigation by boreholes of the kinds and distribution of materials below the land surface at sites of proposed dams and reservoirs is a necessary prerequisite to design and construction. Boreholes are drilled for many purposes (Table 9-1), including recovery and logging of cores of rocks and samples

TABLE 9-1

Objectives of dam-site borehole investigations

I.	Proposed dam, usually an embankment dam, is to be constructed so as to rest partly or entirely on unconsolidated surficial deposits. Borehole investigations are made to ascertain the following, in the field and/or in samples sent to the laboratory.

    A. Vertical and lateral variations in mineral composition and physical properties including descriptions and classification of various units according to a conventional system (such as the *Unified Soil Classification*).

    B. Significant engineering properties including density, shearing strength, compressibility, effects of changes in moisture content, and effects of structural loading.

    C. Vertical and lateral variations in permeability as they may localize undesirable or impermissible subsurface seepage.

    D. Depth to bedrock or to a laterally extensive impermeable layer of unconsolidated material, if a cut-off below the dam is under consideration.

II.   Proposed dam (either a concrete dam or an embankment dam) is to be constructed on bedrock after excavation and removal of unconsolidated surficial deposits and removal of incompetent altered and/or fractured rock in dam foundation. Borehole data obtained from cores and from testing in the field and/or laboratory are made to estimate or measure the following:

    A. Depths and contours on bedrock surface below unconsolidated surficial cover.

    B. Lithology of bedrock units including identification of depths, locations, and attitudes of original planar and linear features in anisotropic rocks, such as bedding planes, foliation and schistosity, and flow structures in igneous rocks, etc.

    C. Location, intensity, and lateral and vertical extent of weathered or otherwise altered zones.

    D. Locations, lateral and vertical extent, orientation and spacing of fractures, including faults and joint surfaces (i.e., smooth, rough, altered, etc.).

    E. Locations and estimation of potential seepage through zones or single channelways of primary or secondary permeability by water testing. Particular care should be exercised in investigations in foundations that contain brittle, fractured rocks and/or soluble rocks such as limestone, dolomite, and gypsum.

    F. Geophysical logs from appropriate borehole probing instruments. Investigation may include estimations of strength as expressed by measured elastic constants.

    G. Estimation or measurement of engineering properties, utilizing cores, of engineering properties, such as density, permeability, effects of water saturation, shear strength, strain characteristics, and response to structural loading.

of unconsolidated deposits, water-testing for permeability, in situ testing of strengths of rocks, and logging by appropriate geophysical devices, of which there is a wide variety. Boreholes are used not only to test foundations, but also serve a useful purpose in obtaining samples and estimating volumes of materials which are to be used in construction of a dam and appurtenant features.

The intensity of investigation by boreholes varies with the level of an investigation. During the reconnaissance and feasibility stages of study holes are drilled after completion of detailed geological mapping and surface geophysical measurements to verify tentative conclusions or to obtain critical information concerning subsurface geological conditions where surface exposures of buried materials are absent, or few in number and/or small in area, and the interpretation of subsurface conditions is impossible or highly uncertain without borehole data. If investigations indicate that construction of a dam and reservoir is feasible, an additional program of drilling is undertaken to provide the data required for safe design of the facilities to be built and to enable prediction of foundation problems that may be encountered during construction.

Borehole data are correlated with surface geologic maps and geophysical data to prepare, to the extent possible, plan maps and vertical sections showing the vertical and horizontal distribution of units of various materials with differing properties. Care should be taken to make a clear distinction between factual data and inferences drawn from these data. Interpolation and extrapolation for long distances should not be substituted for additional borehole testing in arriving at conclusions concerning the geology of bedrock and unconsolidated deposits, particularly where rapid vertical and/or lateral variations are known to exist or are probable.

Boreholes are drilled by a variety of types of equipment including rotary drills with diamond bits or cutting heads, hand or power augers, and churn drills. Auger drills yield highly disturbed samples and are drilled primarily to test unconsolidated materials in borrow areas for possible use in construction or to locate the groundwater table. Churn drills yield cuttings only and are drilled primarily to locate bedrock, the water table, or to penetrate unconsolidated deposits for core drilling in underlying bedrock.

Exploration of foundations in unconsolidated deposits commonly involves the use of rotary drilling rigs and drive-tube samplers which, depending on their design, recover more or less "undisturbed samples" which are then tested for their engineering properties in the field or in the laboratory. A special type of drive sampler, called a "split-tube penetration spoon", yields a core sample and simultaneously permits measurement of resistance to penetration as measured by the number of blows of a standard weight required to produce one foot of penetration. In gravelly and rocky deposits or cohesionless sands and soft wet materials recovery of samples usually requires the use of "piston-type samplers". Water testing in unconsolidated deposits is accomplished by pump-yield tests utilizing perforated casing to support walls of the drill-hole or by forcing water under controlled pressures into intervals of concern of the hole through calibrated, slotted casing.

Core drills with diamond bits are used in exploration of bedrock. Core sizes are conventionally designated as EX, BX, AX, and NX corresponding to diameters of $\frac{7}{8}$, $1\frac{1}{8}$, $1\frac{5}{8}$, and $2\frac{1}{8}$ inches, respectively. Approximate diameters of holes drilled by the bits are $1\frac{1}{2}$, $1\frac{7}{8}$, $2\frac{3}{8}$, and 3 inches, respectively. Rocks that give excellent core recovery with AX or NX bits may break up badly with EX or BX bits, so that it is important to use the largest practicable size in drilling, otherwise a false impression of the quality of the rock may be obtained.

UNCONSOLIDATED DEPOSITS

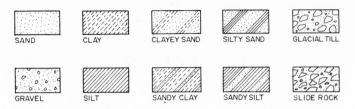

CONSOLIDATED DEPOSITS

Fig.9-6. Graphic symbols for plotting logs of unconsolidated and consolidated (rock) materials.

LOGS OF BOREHOLES

Logs of boreholes are obtained from subsurface investigations of foundations, reservoir areas, and sites for obtaining materials for construction (borrow areas). Data are recorded in tables and graphic plots supplemented with descriptions and interpretations, and commonly are translated into cross-sections or three-dimen-

GEOLOGIC LOG

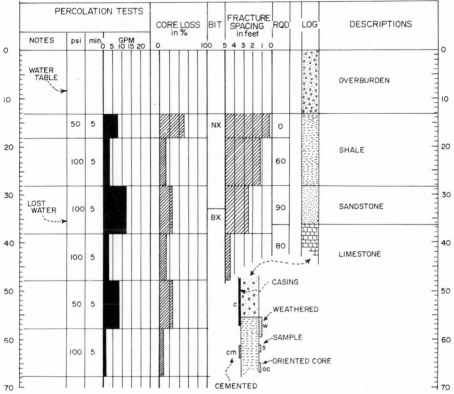

Fig.9-7. Simplified typical log of a borehole in unconsolidated and consolidated (rock) materials. Rocks below a layer of overburden were recovered by diamond-bit coring.

sional plots showing lateral and vertical variations in the properties and distribution of subsurface materials of engineering and geological concern.

There are many types of borehole logs, each serving a particular purpose and obtained from an investigation of a particular kind. Accordingly there is consider-able latitude in the choice of the data that are represented on a log, and it is common practice to design a log form for graphical representation and written notations suited to a particular, specialized purpose.

Fig.9-6 shows symbols ordinarily used in plotting logs of unconsolidated and consolidated natural materials. Fig.9-7 is a simplified typical geological log of a borehole penetrating bedrock below a cover of overburden. In Fig.9-7 the RQD column is a "rock quality designation" which states the percentage of firm, hard pieces of core 4 inches or longer in a given interval of core (Deere, 1968).

An appreciation of the purposes and variety of logs of core holes that are employed in geological and engineering investigations is obtained by review of Table 9-2. A useful guide in obtaining data for preparation of logs in engineering investigations is "Special Procedures for Testing Soil and Rock for Engineering Purposes", Special Technical Publication 479, published by the American Society for Testing Materials.

TABLE 9-2

Types of logs of boreholes

I.   Geological logs from core drilling and percolation tests. Graphical and numerical representation with descriptions and legend for special symbols. Log should include:
     A. Results of percolation tests.
     B. Core loss as a percentage (or core recovery as a percentage).
     C. Core bit size.
     D. Fracture spacing.
     E. RQD rating.
     F. Graphical log of rock types.
     G. Descriptions of rock types.
     H. Special notations for water table, lost drill water, casing, cemented intervals, weathered or altered rock, locations of samples taken for analysis and locations of oriented cores.

II. Geophysical logs from instruments lowered into hole. Method or methods employed are selected on the basis of their particular utility. Among some of the methods available are the following:
     A. Resistivity and microresistivity methods measuring apparent electrical resistivity.
     B. Sonic method measuring velocity of compressional waves as a function of depth.
     C. Formation density methods, using medium energy gamma rays.
     D. Neutron bombardment methods for measurement of porosity.
     E. Gamma-ray methods for identification of rock types by measurement of natural radioactivity.
     F. Methods measuring modulus of elasticity and Poisson's ratio utilizing induced compressional and shear waves.
     G. Methods measuring potential difference between a borehole electrode and a surface electrode.
     H. Temperature measurement methods for determination of thermal gradients.
     I. Methods measuring magnetic properties.

III. Logs of test pits or auger holes, sometimes called "soil profile" logs. Results of drilling and geophysical measurements are translated into cross-sections. Graphical logs include:
     A. Classification symbols.
     B. Size and type of test sample taken.
     C. Descriptions of samples.
     D. Results of size classification by screening or other methods.
     E. Engineering properties.
     F. Elevation of water table, if encountered.

Table 9-2 (continued)

IV.   Logs of cored unconsolidated deposits. Samples obtained by means of drive tubes. Penetration resistance may be determined at same time that drive-tube sample is obtained. Plotted logs contain following entries:
A. Descriptions of methods used to advance hole through successive intervals.
B. Data for calculation of penetration resistance.
C. Moisture content of cores.
D. Description and classification of materials.
E. Graphic log showing materials by symbols.
F. Graphic log showing variations in engineering properties, such as penetration resistance.
G. Graphic representation and/or written summary giving results of percolation tests.
H. Elevation of groundwater table.
I. After laboratory testing of samples, additional entries may be made on log, as appropriate.

V.   Driller's logs provide valuable data on the mechanical details of a drilling operation together with notations by the driller concerning factors effecting the progress of the drilling operation. Frequently drillers include descriptions of materials recovered from holes in their notes. Depending on the nature of the drilling operation, the driller might record in a diary the following kinds of data:
A. Surface equipment type.
B. Equipment used to advance hole.
C. Core size and replacement of worn bits.
D. Drilling time.
E. Loss of drilling water or intersection of groundwater under head.
F. Special operations, such as cementing hole, placing casing, changing course of hole, etc.

FIELD PERMEABILITY TESTS IN BOREHOLES

Water-testing for permeabilities of natural materials penetrated by boreholes is standard procedure and provides the basis for planning necessary measures for overcoming problems created by impermissible seepage beneath dams and around their abutments. The best results are obtained in testing uniformly permeable media such as well-sorted gravels and some sedimentary rocks. The least reliable results, sometimes grossly misleading, come from pressure testing of highly lenticular unconsolidated deposits and rocks containing complex fracture patterns or irregularly interconnecting solution openings, as in limestones. Nevertheless, although it may not be possible to determine or estimate a coefficient of permeability, water-testing will reveal much concerning the possible or probable flow of water through subsurface materials under the hydrostatic head generated by filling of a reservoir.

A variety of procedures is available for permeability measurements, both in the field and in the laboratory, but we are concerned here only with field tests in boreholes. There are two kinds of borehole tests in wide use: (1) bottom-hole tests using the bottom open end of a pipe casing or a length of calibrated perforated casing at the bottom of the casing and (2) packer testing, using either a single packer or two packers a measured distance apart.

In the open-end method either a gravity or a combination gravity-pressure head is employed and a head is determined by calculating the gravity head alone or by combining the calculated gravity head with the gauge pressure head created by pumping at the collar of the hole. Other required parameters are the constant rate of flow of water under a constant head, the internal radius of the casing, and the elevations of the top and bottom of the casing.

To calculate the permeability the following equation is employed:

$$k = Q/5.5rH \tag{9-1}$$

in which k is the permeability, Q the constant rate of flow into the borehole, r the internal radius of the casing, and H the differential head of water in feet. The value of H for gravity-head tests above the groundwater table is the depth of water in the hole. For gravity tests below the groundwater table, H is the difference between the elevation of the groundwater table and the elevation of the water in the hole. For pressure tests the applied pressure is added to the gravity head. If H is expressed in feet, use is made of the relationship one psi = 2.31 ft. For various units of permeability consult Chapter 5. In the United States k commonly is measured in feet per year, Q in gallons per minute, and H in feet, or psi converted to feet.

When a short open-end perforated casing is used at the bottom of a hole a factor is added to the value for r in eq.9-1, which in effect amounts to increasing r by the radius of a circle with a total area equal to the combined areas of the perforations.

Packer tests can be made in unconsolidated materials only when the borehole remains open below the bottom end of the casing. In rocks in which boreholes remain open during water-testing either a single packer or two packers may be employed. In broken rock which may have to be cemented or cased before the borehole can be deepened, the single packer method is preferred.

With the double packer method two packers a fixed distance apart, say 10 or 20 ft, are systematically lowered or raised in the hole to determine permeability in successive intervals by applying water under controlled, fixed pressure at the top of the hole and measuring the flow of water over a period of time, commonly 5 to 10 min. With a single packer an interval of hole is drilled and then immediately tested by placing the packer at the elevation of the bottom of the previously drilled interval or, if necessary, at a slightly higher elevation in firm rock. Not uncommonly broken zones at the bottom of a hole are tested and then cemented before drilling the next interval.

The theory for determining the permeability, k, for a measured interval in a borehole is complex. A widely used formula containing certain simplifications is:

$$k = C_p Q/H \qquad\qquad (9\text{-}2)$$

in which C_p is a factor depending on the length of the test section and the diameter of the borehole. C_p usually is given in tables or is obtained from curves such as those in Fig.9-8. The curves in Fig.9-8 were prepared for use with eq.9-2, when k is

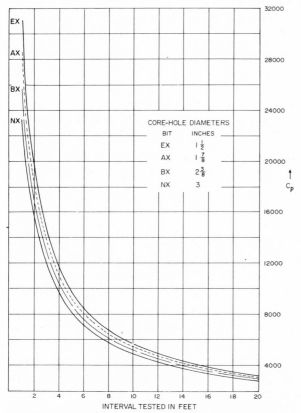

Fig.9-8. Values for C_p in eq.9-2 for various drillhole diameters when k is in feet per year, Q is in gallons per minute, and H is the average head in feet of water in the test interval.

in feet per year, Q is in gallons per minute, and H is the average head in feet of water in the test interval.

When the test interval is below the water table, H is the distance from the water table to the collar of the hole plus the pressure applied at the collar in units of feet of water. Above the water table H is the distance from the middle of the test interval to the collar added to the applied pressure in units of feet of water (one psi = 2.31 ft).

Calculation of permeabilities in rocks in which there is secondary permeability resulting from fractures or solution cavities can be very misleading. For example, the flow of water from a 20-ft test section might be localized along a single open fracture or solution channelway accidentally intersected by the borehole, and the calculated value of k then becomes entirely meaningless when applied to the entire section. Accordingly, in logs of water tests of boreholes in rocks containing secondary permeability it is common practice to record only the pressures and the rate of flow as measured at the collar of the hole.

CORE-HOLE SURVEYS AND ORIENTED CORES

Core holes rarely follow straight lines during drilling, and the precise course followed by a particular hole can be determined only by a survey. Many types of devices are available for hole surveys and enable detection of changes of direction and hole inclination with depth. Some of the most successful devices record azimuth (direction) and inclination on a photographic emulsion.

Planar features in rocks recovered from core-drilling operations usually have unknown strikes and dips. Under many circumstances knowledge of the orientation in bedrock of features such as faults, joints, bedding planes, and foliation planes is of vital importance in interpreting details of subsurface geologic structure and correlating them with features observed at the surface. In modern core-drill exploration of bedrock foundations increasing use is being made of oriented cores from which the strike and dip of planar features can be ascertained.

In a typical operation an oriented core is obtained by drilling one or two to several feet with a core bit. A requirement is that the core does not become detached from the bottom of the hole so as to rotate from its original orientation in the core barrel. Next, a device which photographically records the azimuth and inclination and produces a scribed groove of known orientation on the core is lowered over the undistorted stub of core and the core is brought back to the surface. The scribed groove permits location of the top of the core in horizontal and inclined holes and location of compass directions in cores from vertical holes. In cores from horizontal and inclined holes it is convenient to paint or scribe a reference line on the top of the core after recovery from the hole. The painted or scribed line may be extended for small to large distances along the core by fitting broken segments of core together, depending on the condition of the core after recovery.

The method of determination of true strike and dip of a planar feature in an oriented core from an inclined hole is indicated in Fig.9-9 and 9-10. A scribe mark of known orientation (Fig.9-9) enables location of a vertical plane in the core and the top of the core, which is identified by a painted line. The compass bearing of the

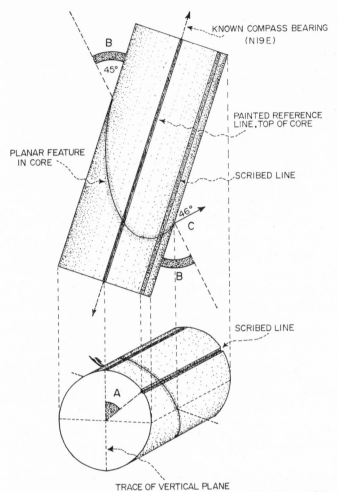

KNOWN COMPASS BEARING
(N19E)

B

45°

PAINTED REFERENCE
LINE, TOP OF CORE

PLANAR FEATURE
IN CORE

SCRIBED LINE

46°

C

B

SCRIBED LINE

A

TRACE OF VERTICAL PLANE

Fig.9-9. Scribed drill core viewed from above and in clinographic projection. See text discussion.

vertical plane and the inclination of the hole are obtained from a photograph and, thus, are known.

The core is placed on a horizontal surface with the painted line on top, and the angle that a planar feature makes with the axis of the core is measured. Also measured is the angle of inclination (dip) of the planar feature, as observed by looking along a horizontal line in the plane of the feature.

The data from the measurements and from the photographic record are plotted on a vellum overlay on a stereographic net as in Fig.9-10. At this stage of plotting the construction shows the compass direction of a vertical plane in the core and the pole for the planar feature when the core is in a horizontal position with the painted line on top of the core.

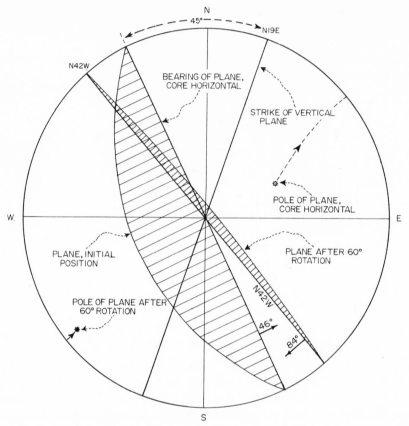

Fig.9-10. Operations with a stereographic net to obtain true strike and dip of a planar feature in the oriented core in Fig.9-8. See text discussion.

It is assumed in this particular example that the inclination of the hole is known to be 60°. To obtain the true strike and dip of the planar feature, the initial plot is appropriately rotated 60° over the net to obtain the location of the pole of the plane in its original orientation in bedrock. A plane 90° from the pole yields the true strike and dip.

Large numbers of measurements can be translated to an equal-area plot for statistical analysis as described in a previous section of this chapter.

BOREHOLE PATTERNS

Investigations of subsurface conditions by boreholes accomplish different purposes at various stages of investigation. Holes are drilled during a preliminary evaluation of a site, during feasibility analysis, and, if the site proves to be feasible

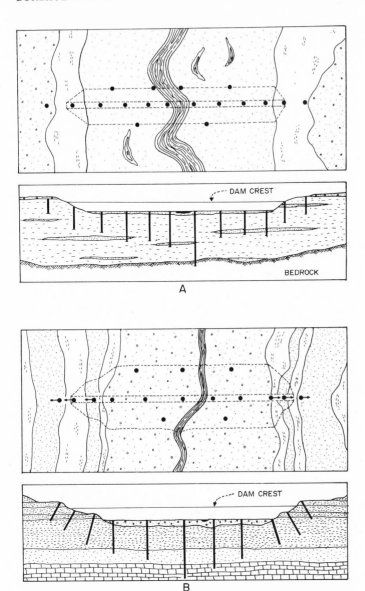

Fig.9-11. Typical borehole patterns at dam sites in plan and section. No scale. (Cont.'d. on p. 210.)
A. Low-profile earth dam on a broad alluvial plain underlain by a thin layer of sand and a great thickness of silty and clayey, unconsolidated sediments. Planned cut-off is sheet-piling.
B. Earth and/or rock-fill dam in horizontal sedimentary rocks. Dam will be constructed on alluvial plain with a cut-off trench extending to a shale layer under valley gravel fill. As a precaution foundation will be grouted from bottom of cut-off trench.

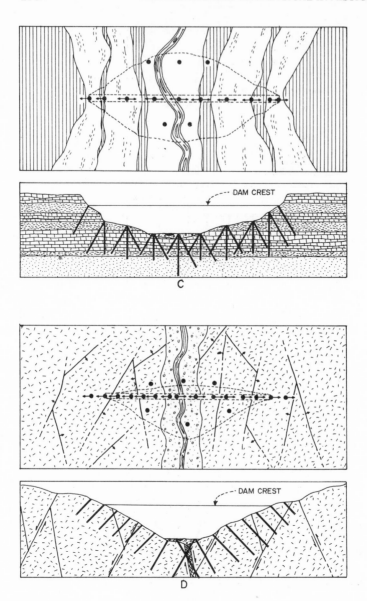

Fig.9-11 (continued).

C. Earth and/or rock-fill dam in horizontal sedimentary sequence known to contain possibly cavernous limestones.

D. Gravity concrete dam in faulted crystalline rocks. Valley was excavated by erosion controlled by a strong fault zone. Minor faults are present in abutments.

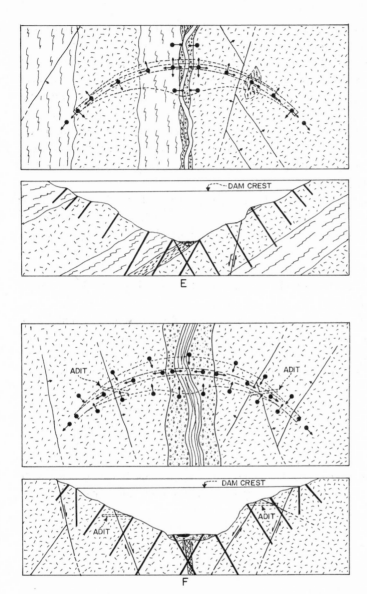

E. Concrete arch dam in inclined layers of granite and metamorphic rocks. Valley appears to have been localized by a soft layer of biotite-rich rock.

F. Concrete arch dam in faulted crystalline rocks. Adits have been driven into abutments for rock-mechanics investigations.

for dam construction, to provide data for the designer of the dam and to assist in the estimation of required quantities and costs for construction.

During the feasibility stage of a dam site the location and approximate dimensions and configuration of the proposed dam and a detailed geologic map should be available so that a program of drilling can be planned that will yield maximum information in proportion to the total footage of drilled holes. A typical borehole pattern contains holes drilled at systematically predetermined locations, off-pattern holes drilled to intersect geologic structures projected downward from the surface geologic map, and additional holes deemed necessary by analysis of data from previously drilled holes.

Although borehole information obtained during the feasibility stage of investigation may lead to the conclusion that economic construction of a dam is not feasible, it should be assumed that data from the holes should be incorporated into a body of knowledge that ultimately will be used by the designing engineer in preparing the specifications for the dam. Accordingly, each borehole and the materials obtained from it should be investigated thoroughly as to their geologic and engineering properties.

During final design stage it is not uncommon for additional boreholes to be drilled to provide needed information to the designer at critical locations in the foundation. If required, boreholes also are located so as to enable engineering analysis of diversion tunnels, underground vaults, and foundations for spillways and powerhouses.

Although each site for a proposed dam possesses its own peculiar topographic and geologic characteristics, planning of systematic borehole investigations of the foundation usually follows well-established procedures. Several examples of borehole patterns in foundations are shown in Fig.9-11. All examples assume relatively simple foundation geology and that the data derived from the boreholes is adequate for safe design of the dam. Boreholes required for study of spillway locations, diversion tunnels, powerhouses, etc., are not shown.

SEISMIC METHODS OF SUBSURFACE INVESTIGATION

The seismic methods of subsurface investigation are particularly applicable in the preliminary and feasibility stages of investigation of dam and reservoir sites. With modern equipment seismic studies not only give quick results, but are relatively inexpensive. Under suitable geologic conditions, that is, when the geometry of shapes and distributions of subsurface units is fairly simple, results of at least 5—10% accuracy are obtained. Normally, seismic surveys, together with electrical resistivity surveys (described in a succeeding section of this chapter) are made after a geologic map has been completed and before testing by boreholes is started.

Seismic data frequently prove to be valuable in planning foundation drilling programs.

Routine investigations yield maximum information in areas of low to moderate relief and have only limited application in steep gorges or canyons. Special types of seismic studies employing boreholes are being given increasing use in measuring elastic constants of rocks.

The most widely used seismic method for foundation exploration is the *refraction method,* in which the times of arrival at a succession of geophones of compressional waves generated by an explosion or by the impact of a falling weight are measured by a precise timing mechanism. The compressional waves, also called *longitudinal waves,* travel with maximum velocities as compared with a variety of other kinds of waves generated by seismic disturbances.

The velocity of transmission of compressional waves through earth materials, V_p, is given by the relation:

$$V_p = [E(1-\nu)/\rho\,(1+\nu)\,(1-2\nu)]^{\frac{1}{2}} \qquad (9\text{-}3)$$

in which E is the modulus of elasticity, ν is Poisson's ratio (commonly with a value near 0.25), and ρ is the density. Clearly, the velocity is especially sensitive to variations in the modulus of elasticity, which, in turn, is a measure of rock strength. The velocity and strength depend on many variables, including rock fabric, mineralogy, and pore water. In general, velocities in crystalline rocks are high to very high. Velocities in sedimentary rocks increase concomitantly with compaction and decrease in pore fluids and with increase in the degree of cementation and recrystallization. Unconsolidated sedimentary accumulations have maximum velocities varying as a function of the mineralogy, the volume of voids, either air-filled or water-filled, and grain size.

Laboratory measurements demonstrate that seismic velocities increase with confining pressures, an observation in accord with the fact that deeply buried rocks have higher velocities than shallow rocks of the same fabric and mineralogy.

Most rocks and unconsolidated deposits are anisotropic. Notable anisotropism is present in stratified sedimentary accumulations and in metamorphic rocks possessing foliation and schistosity. Ordinarily, velocities in the directions parallel to planar structures in anisotropic rocks are greater than in directions perpendicular to these structures.

Table 9-3 lists ranges of velocities for compressional waves in a variety of natural materials. Differences in velocities are expressions of the factors noted above.

Passage of seismic impulses from one medium to another obeys Snell's law, which states that:

$$n = \sin i/\sin r = V_1/V_2 \qquad (9\text{-}4)$$

TABLE 9-3

Velocities of compressional waves

	V_p (ft/sec)	V_p (km/sec)
Igneous rocks		
Basalt	16,500–21,000	5.1–6.4
Diabase	19,000–21,500	5.8–6.6
Gabbro	13,000–22,000	6.5–6.7
Granite	18,000–20,000	5.5–6.1
Metamorphic rocks		
Gneiss	11,500–23,000	3.5–7.0
Marble	12,000–22,500	3.7–6.9
Quartzite	18,000–20,000	5.6–6.1
Schist	11,500–18,000	3.5–5.7
Slate	11,500–17,500	3.5–5.4
Sedimentary rocks		
Dolostone (dolomite)	11,500–22,500	3.5–6.9
Gypsum	6,500–11,500	2.0–3.5
Limestone	9,000–23,000	2.8–7.0
Sandstone	4,500–14,500	1.4–4.4
Shale	6,800–14,500	2.1–4.4
Unconsolidated deposits		
Alluvium	1,000– 2,000	0.3–0.6
Clay (wet)	4,900– 6,500	1.5–2.0
Clay (sandy)	6,000– 8,000	2.0–2.4
Water (25°C)	4,950	1.509

in which n is the index of refraction, i is the angle of incidence in medium of velocity V_1, and r is the angle of refraction in medium of velocity V_2. Both i and r are measured from a normal to the contact between two media of different velocities.

Paths of seismic impulses with several angles of incidence passing from one medium to another are shown in Fig.9-12. For a critical angle of incidence, i_c, the angle of refraction in the medium of velocity, V_2, is 90°, and:

$$n = \sin i_c / \sin r = \sin i_c / \sin 90° = \sin i_c \qquad (9\text{-}5)$$

Impulses following paths with angles of incidence less than i_c are partly reflected and partly refracted, and impulses along paths with angles greater than i_c are totally reflected. The energy traveling along the contact between the two media is a secondary source of impulses which are refracted back into medium of velocity V_1 at the critical angle, i_c.

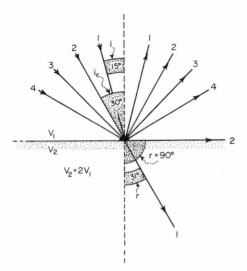

Fig.9-12. Reflection and refraction of seismic impulses at the contact between two media with different velocities of propagation.

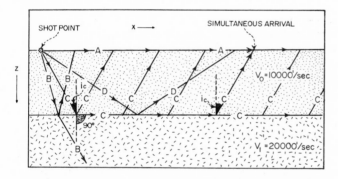

Fig.9-13. Several possible paths followed by an instantaneous seismic impulse generated by an explosion at a shot point. A = energy travels parallel to air-medium contact; B = energy is partly reflected back into medium of velocity, V_0, and is partly refracted into medium of velocity V_1; C = energy travels parallel to contact. Critical angle is i_c; D = energy is totally reflected at contact between two media.

Fig.9-13 shows a few of many possible paths followed by seismic impulses generated by an explosion at a shot point. Because $V_1 > V_0$ an impulse following paths C arrives at a certain distance from the shot point simultaneously with an impulse generated at the same instant and traveling path A.

The application of the refraction method to the measurement of the thicknesses and velocity of propagation of longitudinal seismic waves in horizontal layers is indicated in Fig.9-14. Geophones pick up seismic disturbances generated by an explosion or a blow from a falling weight at the shot point. Although a variety of waves with differing velocities arrive at the geophones, the first impulses to arrive

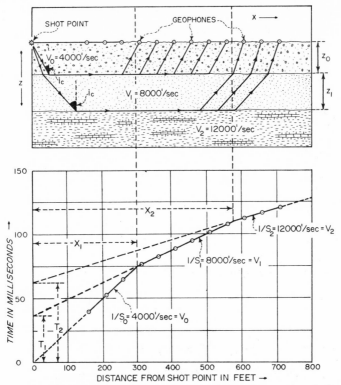

Fig.9-14. Shot point and geophone layout and time-distance plot obtained from a seismometer. See text discussion.

are those of longitudinal waves. Their times of arrival are measured and recorded by a sensitive timing device at a receiver. The times of arrival as a function of distance from the shot point enables construction of a *time–distance plot,* as in Fig.9-14.

In Fig.9-14 V_0, V_1, and V_2 are velocities in a succession of horizontal layers and $V_0 < V_1 < V_2$. Critical angles between media in contact are designated by i_c. Energy from the shot point travels horizontally along the air–medium contact with velocity V_0 and produces a time distance line with a characteristic slope, S_0, such that $1/S_0 = V_0$. The refracted waves in media V_1 and V_2 similarly produce time–distance lines with slopes such that $1/S_1$ (medium V_1) = V_1 and $1/S_2$ (medium V_2) = V_2. Projection of the line segments to the time coordinate yields intercepts $T_0 = 0, T_1$, and T_2.

A useful general equation for multiple horizontal layers numbered 0, 1, 2, . . . n with thicknesses $Z_0, Z_1, Z_2 . . . Z_n$ and velocites $V_0, V_1, V_2 . . . V_n$, is:

$$t_n = X/V_n + 2Z_0 (V_n^2 - V_0^2)^{\frac{1}{2}}/V_0 V_n + 2Z_1 (V_n^2 - V_1^2)^{\frac{1}{2}}/V_1 V_n + $$
$$2Z_2 (V_n^2 - V_2^2)^{\frac{1}{2}}/V_2 V_n . . . 2Z_{n-1} (V_n^2 - V_{n-1}^2)^{\frac{1}{2}}/V_n V_{n-1} \qquad (9\text{-}6)$$

in which t_n is the time that it takes for a refracted impulse penetrating as deep as the nth layer to reach the surface at a distance X from the shot point.

In the single layer case eq.9-6 reduces to:

$$t_1 = X/V_1 + 2Z_0 (V_1{}^2 - V_0{}^2)^{\frac{1}{2}}/V_0 V_1 \tag{9-7}$$

The intercept on the time coordinate, T_1, obtained by projection of a time-distance plot, is:

$$T_1 = 2Z_0 (V_1{}^2 - V_0{}^2)^{\frac{1}{2}}/V_0 V_1 \tag{9-8}$$

For the two-layer case (Fig.9-14) the general equation (9-6) becomes:

$$t_2 = X/V_2 + 2Z_0(V_2{}^2 - V_0{}^2)^{\frac{1}{2}}/V_0 V_2 + 2Z_1(V_2{}^2 - V_1{}^2)^{\frac{1}{2}}/V_1 V_2 \tag{9-9}$$

T_2, the intercept on the time axis (Fig.9-14) is:

$$T_2 = 2Z_0(V_2{}^2 - V_0{}^2)^{\frac{1}{2}}/V_0 V_2 + 2Z_1(V_2{}^2 - V_1{}^2)^{\frac{1}{2}}/V_1 V_2 \tag{9-10}$$

Because time–distance plots enable estimation of $T_0, T_1 \ldots T_n, t_0, t_1 \ldots t_n, X_0,$ $X_1 \ldots X_n$, and the velocities, $V_0, V_1 \ldots V_n$ can be calculated as the reciprocals of appropriate line segments, information is available that permits calculation of $Z_0,$ $Z_1 \ldots Z_n$ by eq.9-5–9-10, above.

In the single layer case either 9-7 or 9-8 is utilized. The simpler relationship employing only the intercept time is expressed in eq.9-8, which, rearranged, is:

$$Z_0 = T_1(V_0 V_1)/2(V_1{}^2 - V_0{}^2)^{\frac{1}{2}} \tag{9-11}$$

If desired, eq.9-7 may be used to check the results obtained from eq.9-11.

In the two-layer case (Fig.9-14), we may use the intercept time, T_2, and by rearranging eq.9-10 obtain:

$$Z_1 = V_1 V_2 [T_2 - 2Z_0(V_2{}^2 - V_0{}^2)^{\frac{1}{2}}]/2V_0 V_1(V_2{}^2 - V_1{}^2)^{\frac{1}{2}} \tag{9-12}$$

Alternately, and to check the results from calculations with eq.9-12, use is made of eq.9-9.

In general practice, the above calculations are greatly expedited by employment of computer programs.

The refraction method also is used for determining the angles of slopes of buried surfaces, as, for example, the angles of slopes of bedrock surfaces beneath floodplain deposits. In Fig.9-15 α is the apparent slope of a buried surface in the

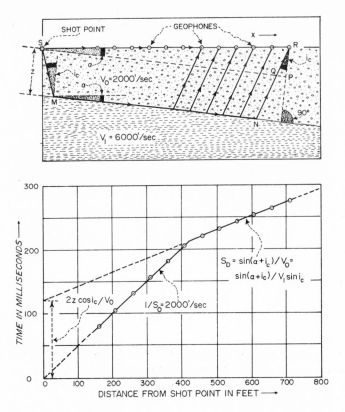

Fig.9-15. Down-slope seismic traverse above an inclined buried surface. See text discussion.

vertical plane including a downslope seismic traverse. The travel time from S to R via $SMNQR$ is:

$$t_D = X(\cos\alpha - \sin\alpha \tan i_c)/V_1 + 2Z \cos i_c/V_0 + X \sin\alpha/V_0 \cos i_c$$

$$= X \sin (\alpha + i_c)/V_0 + 2Z \cos i_c/V_0 \qquad (9\text{-}13)$$

The time–distance plot has a slope, S_D, of:

$$S_D = \sin (\alpha + i_c)/V_0 = \sin (\alpha + i_c)/V_1 \sin i_c \qquad (9\text{-}14)$$

and an intercept on the time axis of:

$$T_D = 2Z \cos i_c/V_0 \qquad (9\text{-}15)$$

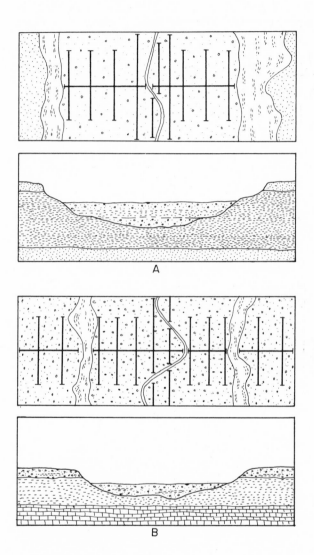

Fig.9-16. Idealized seismic traverse layouts in plan and interpretations in cross-section derived from time–distance curves and some knowledge of local general geology. (Cont.'d. on p. 220.)
A. Stratified valley fill lies above shale bedrock.
B. Alluvium is present beneath terraces and below a floodplain. A limestone layer is present at depth.

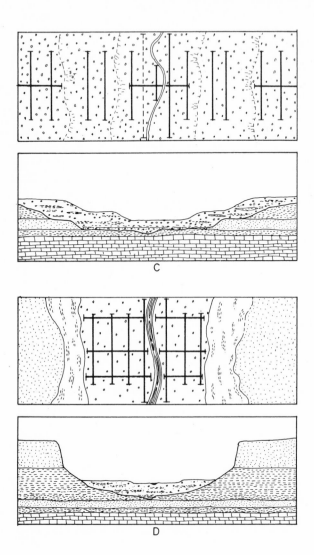

Fig.9-16 (continued).
C. Terraces in a broad valley are underlain by alluvial deposits resting on a succession of terraces in bedrock. A limestone is present at depth.
D. Alluvium rests on a shale layer immediately above a sandstone layer.

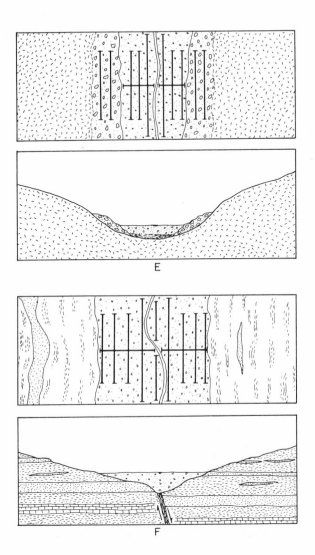

E. Glacial till in lateral moraines and fluvioglacial deposits cover the floor of a glaciated valley in crystalline rocks.

F. Seismic investigations indicate the presence of a fault in bedrock beneath the valley fill.

In an upslope seismic traverse using the same shot point the equations are:

$$t_u = X \sin{(i_c - \alpha)}/V_0 + 2 Z \cos{i_c}/V_0 \tag{9-16}$$

$$S_u = \sin{(i_c - \alpha)}/V_0 = \sin{(i_c - \alpha)}/V_1 \sin{i_c} \tag{9-17}$$

From the above equations we obtain:

$$i_c = \tfrac{1}{2} \, [\sin^{-1}{(V_0 S_D)} + \sin^{-1}{(V_0 S_u)}] \tag{9-18}$$

$$\alpha = \tfrac{1}{2} \, [\sin^{-1}{(V_0 S_D)} - \sin^{-1}{(V_0 S_u)}] \tag{9-19}$$

A further useful relationship is:

$$V_1 = 2 \cos\alpha/(S_u + S_D) \tag{9-20a}$$

or, approximately:

$$V_1 \approx 2/(S_u + S_D) \tag{9-20b}$$

Existence of a sloping buried surface usually is suggested by dissimilar time–distance plots in opposing directions along a straight-line seismic traverse. The angle of slope, α, as calculated from measurements may be only an apparent angle of slope. To obtain the full angle of slope traverses are run in mutually perpendicular directions. Commonly traverses at right angles to the trend of a valley yield full slope angles of buried surfaces.

Several idealized seismic traverse layouts and the interpretation based on time–distance plots derived from them are shown in Fig.9-16. In areas of unknown stratigraphy identification of the lithologies of the various layers is not certain. Instead, the layers are identified by their characteristic velocities, and tentative interpretations are made of the various rock types subject to later verification by core drilling.

RESISTIVITY GEOPHYSICAL INVESTIGATIONS

Investigations with electrical-resistivity geophysical equipment have great utility in interpretation of subsurface geological conditions at dam sites, in reservoir basins, and in borrow areas for construction materials. Used in conjunction with seismic surveys and borehole-data resistivity investigations supplement and complement data derived from the other techniques so as to increase confidence in construction of a three-dimensional picture of the kinds and distributions of consolidated and unconsolidated deposits at shallow to moderate depths beneath the earth's surface. Resistivity surveys have been used successfully in rapid, low-cost reconnaissance studies of sand and gravel deposits, floodplain deposits, layered sedimentary rocks, groundwater table, and quarry rock.

The resistivity method is based on measurement of the *apparent resistivity* between two voltage (potential) electrodes in an electric field generated by an electrical current produced at two current electrodes (Fig.9-17). There are several

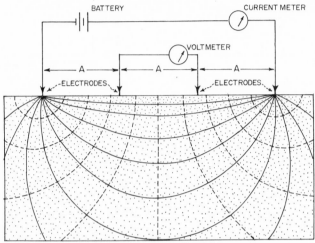

Fig.9-17. Wenner configuration for measurement of apparent resistivity. Current flow lines and equipotential lines (dashed) generated at current electrodes are shown for a material of uniform resistance and infinite depth.

possible configurations of the electrodes, but the one that has been used most extensively in engineering investigations is the *Wenner configuration* (Wenner, 1915) in which the spacing between the electrodes is equal.

In Fig.9-17 the elements of the Wenner method and the electrical field generated by two current electrodes are shown diagrammatically for an assumed homogeneous medium of infinite depth. If the spacing between the electrodes is A, the apparent resistivity between the voltage electrodes, ρ is:

$$\rho = 2\pi A(V/I) = 2\pi AR \qquad (9-22)$$

in which V is the voltage between the voltage electrodes, I is the current introduced at the current electrodes, and, by Ohm's law, $R = VI$, is the electrical resistance of the material as expressed in standard units (ohm-cm or ohm-ft; 1 ohm-cm = 0.0328 ohm-ft). Some representative values for resistances of natural materials are given in Table 9-4.

There are two generally employed methods for making investigations of subsurface conditions employing the Wenner configuration. In one method a fixed electrode spacing, predetermined by experimentation in the field, is used in making

TABLE 9-4

Resistances of some common materials

Material	Resistance (ohm-ft)*
Hard bedrock. Dry sand and gravel.	8000 and more
Slightly fractured bedrock. Sand and gravel with layers of silt.	1000–9000
Silty sand and gravel	1000, approx.
Fractured bedrock with water-filled cracks	500–1000
Wet, silty clay	10– 50
Saturated or moist clay	5– 10

*To convert ohm-ft to ohm-cm divide by 30.48.

a linear traverse by "leap-frogging" the electrodes. The apparent resistivities, as calculated by eq.9-22 are plotted against distance and show variations in resistivities dependent on variations in natural resistivities of materials at depth. Commonly, an electrode separation is used which is of the same order of magnitude as the desired depth of significant current penetration.

An hypothetical example showing the results of a resistivity traverse with fixed electrode spacing is shown in Fig.9-18. It is assumed that a fault zone in crystalline rocks beneath a gravel layer is filled with electrolyte-bearing water and has a lower resistivity than the surrounding rocks and the gravel layer above. Similarly, a wet

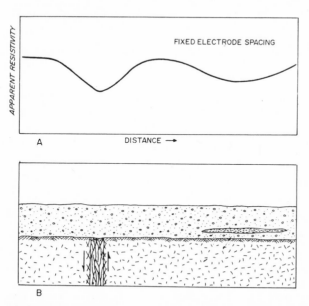

Fig.9-18. Hypothetical apparent resistivity curves obtained from a linear traverse with a fixed electrode spacing over a water-bearing fault beneath a gravel stratum and a clay lens in the gravel stratum.

clay lens in the gravel layer has a lower resistivity than the surrounding gravel. The plot of resistivity against distance reveals the existence of the low-resistivity zones but does not distinguish between them. Correct interpretation of the resistivity curve can be made only with knowledge derived from other sources such as boreholes or, possibly, seismic surveys.

A second method of exploration with the Wenner configuration is based on a well-established empirical relationship in which it states that the depth of investigation is approximately equal to the electrode spacing. In this method the electrode spacing at a particular location is increased systematically, and apparent resistivities are plotted against electrode spacing using ordinary rectangular (Cartesian) coordinates or two-cycle logarithmic graph paper.

The basis for the variable electrode spacing method (Wenner configuration) is indicated in Fig.9-19. In Fig.9-19A a layer of resistivity, ρ_1, lies over a layer of higher resistivity, ρ_2. For a sufficiently large electrode spacing the current penetrates the contact between the two layers in a significant amount, a fact which is reflected in the readings from the voltmeter between the potential electrodes. With a close electrode spacing the resistivity measured at the surface is essentially that of the upper layer because most of the current flow is confined to this layer.

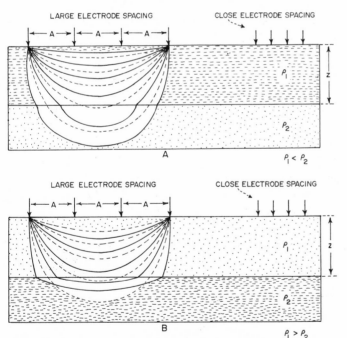

Fig.9-19. Flow of current between current electrodes in a two-layer model. Dashed lines indicate current flow in an unlayered, uniform material.
A. A layer of relatively low resistivity lies over a layer of higher resistivity.
B. A layer of relatively high resistivity lies over a layer of lower resistivity.

Lines of current flow for a large electrode spacing when a layer of high resistivity lies above a layer of lower resistivity are indicated in Fig.9-19B. In both representations in Fig.9-19 the locations of current flow lines as they would exist in an unlayered, homogeneous medium are shown by dashed lines.

Fig.9-20 diagrammatically illustrates the curves of resistivity plotted against

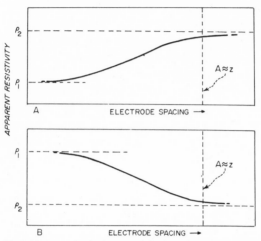

Fig.9-20. Hypothetical curves of resistivity plotted against electrode spacing for the two-layer models in Fig.9-19. The electrode spacing, A, which might be assumed to equal the depth, z, of the contact between the two layers is indicated. (See text discussion.)
A. A layer of relatively low resistivity lies above a layer of higher resistivity.
B. A layer of relatively high resistivity lies above a layer of lower resistivity.

electrode spacing for the two situations depicted in Fig.9-19. It will be noted that as the electrode spacing is increased, the measured resistivity approaches that of the lower layer. Under many circumstances the depth to the contact very closely approximates the electrode spacing, as corroborated by seismic investigations and/ or boreholes. In practice, correlation of electrode spacing with depth is not attempted until field experimentation where depths are actually known supplies a basis for quantitative interpretation of depths from resistivity data.

Interpretation of resistivity data under some circumstances is assisted by methods other than simply plotting apparent resistivities against electrode spacings. Two methods that have been extensively utilized are the "Barnes Layer Method" (Barnes, 1945) and the "Moore Cumulative Method" (Moore, 1945). The interested reader is urged to review these references and an extensive literature in the field of applied geophysics.

ROCK- AND SOIL-MECHANICS INVESTIGATIONS

Rock- and soil-mechanics investigations of cores and drive samples from boreholes and by instrumentation of boreholes and, possibly, adits or shafts, are necessary operations in modern planning and design of dams, reservoirs, and appurtenant features, such as tunnels, spillways, and powerhouses. The primary objective of these investigations is the determination of the expected behavior of subsurface materials during excavation and after loading by surface facilities. Data providing a basis for predicting the interaction between foundation rocks, or unconsolidated deposits, and superimposed loads are required by the designing engineer in preparation of plans and specifications with a generous margin of safety and rock mechanics measurements, together with geological, geophysical, and borehole information, must be carried to a level adequate for the purpose.

EARTHQUAKE HAZARDS

Repeated, disastrous destruction of life and property by major earthquakes throughout history is recorded in the annals of many countries. Most earthquakes originate in well-defined seismic belts in regions of persistently active crustal movements. About 80% of the world's earthquakes are generated in a circum-Pacific seismic belt, which coincides with the Pacific "belt of fire", so called because of the existence of active and recently active volcanoes. More restricted seismic belts are present around the Mediterranean Sea and the Indian Ocean and in other parts of the world.

Locations of epicenters of earthquakes in seismic belts as well as in areas of infrequent activity are plotted annually on world maps distributed by the National Earthquake Information Center of the National Oceanic and Atmospheric Administration (NOAA) at Boulder, Colorado. The locations of the earthquakes are determined by NOAA environmental research laboratories and from data furnished by many cooperating foreign and domestic seismological stations.

Although by far the greater proportion of earthquakes is confined to well-defined seismic belts near continental margins, destructive earth tremors have jolted areas that were considered as aseismic because of their large distances from zones of concentrated, continued seismic activity. In the United States notable earthquakes at a considerable distance from any known seismic belt caused extensive damage at New Madrid, Missouri, in 1811, and at Charlestown, South Carolina in 1886. Thus, it is clear that no segment of the earth's surface can be considered as totally safe from destructive earth tremors, although in many areas the statistical probability of destruction of life and structures by earthquakes may appear to be remotely small.

Some earth tremors are associated with volcanic activity, but earthquakes of destructive force almost invariably are caused by sudden release of elastically stored energy along faults in shallow and deeper portions of the earth's crust. Along some faults dislocations that generate earthquakes reach the earth's surface and cause visible displacements. However, most earthquake-causing faults generate seismic energy by movements at depth, although subsequent exposure by erosion may reveal the sites of dislocations of rocks in the geologic past.

Shock waves associated with earthquakes are complex but can be classified into three general types. The type of wave which moves with maximum velocity is the *P wave*, also called a *longitudinal wave*. *S waves* are *shear waves* which at shallow depths are propagated with a velocity approximately 60% of that of P waves. P and S waves are *body waves* that travel through rocks *below* the surface. L waves are relatively slow surface waves of long periods that are capable of causing swaying of buildings or wave motion in water bodies at great distances from the point of origin. Most damage from earthquakes is caused by L waves rather than by the P and S waves.

Earthquakes are rated by two scales: the *Mercalli scale of intensity* and the *Richter scale of magnitude*. Richter magnitude is calculated from the amplitudes of ground vibrations observed on an instrumentally recorded seismogram and usually can be determined within a few minutes after arrival at a recording station. The Richter scale measures the size of an earthquake at its source, which, if it is hundreds or thousands of miles away, is of small concern from an engineering point of view. The Mercalli scale is based on observations of actual earthquake effects at a specific location. Because of wide variations in intensity within an area as a function of variations in surface and subsurface geologic structure, the emphasis is on the effects observed at a particular point of reference, rather than at the origin.

The Mercalli scale rates intensity by numbers from 1 to 12. At the lower end of the scale earthquake tremors are not felt by people except under especially favorable circumstances, whereas at intensity 12 damage is extreme or total, and rock and soil bodies are visibly and sometimes extensively and rapidly displaced.

The intensity of an earthquake at a particular location is closely dependent on subsurface geologic conditions as expressed by the kinds and distributions of rocks and unconsolidated deposits. Except for structures built directly above a fault zone, proximity to a fault may be of less consequence than the subsurface geologic structure. Damage to structures built on solid rock, even close to an epicenter, usually is much less than damage to structures resting on soft, unconsolidated material, especially if the material is water-soaked (Iacopi, 1971).

Dams built of concrete and resting on solid bedrock usually are not extensively damaged by an earthquake. Well-designed and constructed earth or rock-fill dams, although dislocated more than concrete dams by shock waves, generally have the capacity to adjust to minor displacements without failure, assuming, of course, that they are not built on very weak foundation materials or are not dislocated by extension of fault movements into the dam.

Of particular concern, although visible damage may not develop in a dam as a consequence of an earthquake, is the possibility that small-scale displacements in foundation rocks may alter the pattern of flow of groundwater beneath the dam, and in places disrupt or reopen cracks filled and sealed by grouting operations. Within the reservoir site earthquakes of small to great intensity may promote extensive landsliding in slopes underlain by precariously perched bodies of unconsolidated, frequently water-soaked unconsolidated deposits, or in masses of rock in unstable over-steepened slopes.

Clearly, proposed construction of a dam and reservoir in an area of moderate to high earthquake incidence requires a very careful and thorough study of all surface and subsurface geologic features during the feasibility stages of investigation, so that a determination can be made that construction of the dam will not present a continuing threat to life and property in the flow channel below the dam. Even in supposedly aseismic areas the added cost of designing and building structures so that they will resist damage by earthquakes can be justified by the fact that incorporation of earthquake-resistant features also improves the short and long term safety of a structure from other points of view. In any event, provision in design and construction for measures tending to nullify the effects of natural disasters of all kinds, including destructive earthquakes, should be considered as social responsibility, even though historical records do not appear to justify incorporation of such measures.

Because of the danger of failure of unstable slopes as a consequence of earthquake tremors, particular care should be given to the locations and volumes of materials in potentially dangerous slide areas in reservoirs. If possible collapse of slopes presents a threat of destructive overflow of a dam, steps should be taken to stabilize the slopes by suitably appropriate means.

CONSTRUCTION-STAGE GEOTECHNICAL INVESTIGATIONS

Preparation of the foundation for construction of a dam presents an opportunity that will never be realized again to examine and map geologic details that are useful in controlling blanket and curtain grouting and in other improvements of the competency of the foundation materials to support the load of the dam and the water impounded behind it. Although investigations prior to design and construction may have been conducted with meticulous care, practically every foundation excavation reveals unanticipated conditions that require immediate attention, and, in some instances, revision of estimated quantities and details of design.

A carefully prepared current geologic map, and a complete photographic record of progress of construction are essential in assessing the nature of problems encountered during foundation excavation and in determining procedures for rectifying or improving undesirable conditions.

Methods of overcoming or improving the competency of dam foundations during construction are reviewed in the next chapter.

POST-CONSTRUCTION GEOTECHNICAL INVESTIGATIONS

Data concerning the long-term changes in foundations owing to percolation of groundwater and slow responses to loading by a dam and water stored behind it are practically nonexistent. This is so because most of the great and many of the small dams now in existence have been constructed only within this century. The sciences of rock and soil mechanics are relatively young, and the future holds bright hopes for their increasing utility in the design and construction of dams and reservoirs with adequate provision for safety against gross and subtle changes that accompany aging through spans of time far beyond our present experience, but many dams now existing were built without the benefit of recent advances in technology.

Whereas a dam and a reservoir are surface features that are subject to constant inspection and correction of defects as they appear, rocks or soils in foundations are buried beneath the dam or stored water above them, and changes in their properties with time can not be observed directly.

Instrumentation of dams for observation of internal changes with time and changing loads produced by a fluctuating reservoir level has reached a high state of sophistication and is routine. Because the structural integrity of a dam can be no greater than that of its foundation, increasing attention should be given to instrumentation of foundations in such a manner that changes in conditions, whether slight or not, can be monitored continuously during the expected lifetime of the dam and reservoir.

Of particular concern are slow, quasiplastic adjustments to load over extended periods of time and physical and chemical changes in rocks and unconsolidated deposits, owing to groundwater seepage beneath and in the abutments of the dam. As technology and the quality of instrumentation advance, there is abundant justification for periodic reassessment of the foundation geology and subsurface hydrology by geophysical, rock mechanics, and borehole investigations and installation of modern, improved monitoring devices which will foretell the need for corrective measures well in advance of possible foundation disintegration and failure.

REFERENCES

Barnes, H. E., 1954. Electrical subsurface exploration simplified. *Roads and Streets,* April Issue, pp.81–114.

Deere, D. U., 1968. Geological considerations. In: K. G. Stagg and O. C. Zienkiewicz (Editors), *Rock Mechanics in Engineering Practice.* Wiley, New York, N.Y., pp.1–20.

Iacopi, R., 1971. *Earthquake Country.* Lane Books, Menlo Park, Calif., 160pp.

Moore, R. W., 1945. An empirical method of interpretation of earth resistivity measurements, *Trans. Am. Inst. Min. Metall. Engrs.,* 164: 197–215.

Wenner, F. C., 1915. *A Method of Measuring Earth Resistivity.* U.S. Natl. Bur. Standards, Sci. Pap., 258.

Chapter 10

GEOTECHNICAL ASPECTS OF DAM AND RESERVOIR CONSTRUCTION

INTRODUCTION

Construction of a dam and appurtenant features and, if required, improvement of the reservoir site behind the dam terminate and bring into sharp focus all of the efforts expended in preliminary field and laboratory investigations, design, and estimates of quantities and costs. In the foundation and abutments removal of loose surface cover or soil, excavation for a keyway or a cutoff trench in rock or unconsolidated deposits, or removal of weathered or fractured rock produce bare continuous exposures that reveal details of exact location and physical characteristics of subsurface materials that may or may not have been anticipated qualitatively or quantitatively in preliminary investigations, however carefully they were conducted. Problems, whether anticipated or not, now become field problems of immediate urgency and require practical solutions involving men, equipment, time schedules, and conventional or at times, highly innovative procedures.

Countless printed words record construction procedures and special measures that have been employed to overcome problems in dams and reservoirs now existing, and serve to emphasize the fact that nowhere in the world are there two identical dam and reservoir sites. As in many kinds of heavy construction, persons with wide experience have come to expect the unexpected for the simple reason that natural phenomena commonly are not subject to easy preliminary assessment or categorization by arbitrary parameters established by geotechnologists and engineers.

Construction of a dam, in addition to foundation preparations, usually involves a variety of preliminary operations including building of access roads, processing of materials for construction, and construction of diversion of features such as coffer dams and/or tunnels, or surface conduits. The special problems and techniques associated with each of these operations are not reviewed here. Instead, attention will be confined to the field treatment of foundations and abutments to improve their competency and to reduce or eliminate subsurface seepage. Inasmuch as tunnels and underground vaults in rock foundations and abutments may influence total behavior under loading, the reader may find it useful to refer to a summary of the theory and practice of tunneling in rocks by the author (Wahlstrom, 1973).

FOUNDATION EXCAVATION

Based on preliminary investigations a planned program of foundation excavation is initiated with the expectation that the volume of excavated material and the
configuration of the excavation will reasonably approximate predictions or
estimates in the plans and specifications. Generally it is the responsibility of the
construction engineer to establish slopes for excavations that will be permanently
stable or that will not fail during construction. In earth materials slopes of $1-\frac{1}{2}:1$ to
2:1 are excavated in permanent cuts and slopes of 1:1 (45°) are established in
temporary cuts, except where conditions of unusual instability are anticipated. In
bedrock that is not closely fractured or does not contain inclined planes of
potential slippage, such as bedding planes in weak rocks, slopes are excavated at
angles up to the vertical.

In foundations in unconsolidated natural deposits excavation may reveal
inadequate localized or widespread foundation materials that require special treatment or total removal. Unacceptable or inadequate in foundations are unconsolidated materials rich in organic substances such as topsoil, swamp muck, or peat,
loose deposits of sand or silt, slide-rock and talus accumulations, and plastic, active,
sensitive, or swelling clays. Poor foundation conditions in rocks are associated with
close fracturing, weathering or hydrothermal alteration, or poorly indurated sedimentary rocks.

Excavations in bedrock should, to the extent possible, extend into firm, fresh
rock. Closely fractured zones extending downward, especially if containing soft,
altered materials such as clay gouge or products of weathering, should be removed
to the extent feasible. The objective of excavation is preparation of a clean surface
that will provide an optimum contact with the dam materials, whether earth or
concrete, that will be placed on it. Where bearing strength is a factor to be
considered, treatment by grouting or by installation of rock bolts or steel cable
anchorage may be required. Methods for improving strength characteristics of
foundation rocks are described in subsequent sections of this chapter.

Prolonged exposure of both earth and rock foundations to the atmosphere or
to accumulations of water frequently results in deterioration by hydration, dehydration, frost action, or skin shrinkage and expansion with changes in temperature.
It is good practice to protect reactive surfaces that will be exposed for prolonged
intervals of time with gunite or bituminous materials. Alternatively, original cover is
not removed until final cleanup and just prior to placement of the dam.

DAM CONSTRUCTION ON UNCONSOLIDATED DEPOSITS

Ideally, excavations in unconsolidated deposits for a dam should extend to
solid bedrock for the full width of the dam, whether it is constructed of concrete or

of earth and/or rock fill. However, there are many locations where the depth of the valley fill is so great that dams must be constructed in part or entirely on unconsolidated deposits, and, as required, appropriate steps are taken to improve the engineering properties of foundation materials and to reduce subsurface seepage to permissible levels. Except for low dams of small gross weight, concrete dams are not built on unconsolidated deposits because of their generally low-bearing strength. Larger dams constructed in whole or in part on unconsolidated deposits should, without exception, be earth or rock-fill dams with built-in capabilities of adjusting to settlement in foundation materials.

Cross-sections of several earth and/or rock-fill dams constructed at least in part on unconsolidated subsurface deposits are shown in Fig.10-1. The sections show various measures that are taken to eliminate or greatly reduce potential seepage beneath the dams under a variety of circumstances. It is clear that considerable information as to the distribution and permeabilities of subsurface materials is required prior to the design and construction of cut-off features.

GROUTING OPERATIONS IN BEDROCK

The goal of foundation and abutment grouting in bedrock is improvement of strength and bearing capacity and the filling with grout of underground channelways that have a potential for impermissible seepage. Under some circumstances, as discussed in a later section of this chapter, rock bolts and/or steel cables are used to improve the strength of bedrocks, but the most useful technique of general application utilizes drilling and pressure grouting, either with water—cement mixtures or with other types of sealants.

Preliminary geological and geophysical investigations usually reveal the general characteristics of bedrock in the foundation and abutments and enable identification of zones of potential seepage. However, many small but important details of the geology may not be revealed until the keyway for the dam has been excavated and all loose materials have been removed so that a clean bedrock surface can be examined. This is a critical time in dam construction because the constructor is usually eager to proceed with building the dam and does not welcome any delay even though examination of bedrock may indicate the necessity for extensive, time-consuming treatment of the foundation to prevent undesirable, even dangerous water flows. However, it cannot be overemphasized that never again in the life-time of the dam will it be possible to examine in detail and take the appropriate necessary steps to correct adverse conditions that are revealed in a bare bedrock surface, and, considering the vital importance of taking adequate corrective measures *before* the dam is built, and the reservoir is filled, construction schedules should be given only secondary consideration when extensive foundation treatment is required to assure the ultimate safety of the dam.

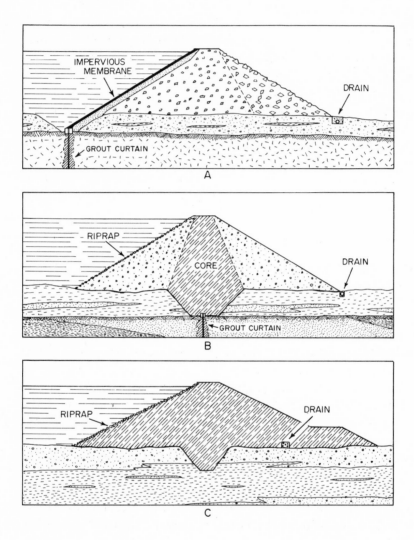

Fig.10-1. Cross-sections of earth and rock-fill dams on unconsolidated deposits.
A. Rock-fill dam. Impervious membrane (asphaltic concrete) extends to a grout cap on bedrock.
B. Cut-off trench extends to bedrock.
C. Cut-off trench penetrates impervious layer in unconsolidated valley fill.

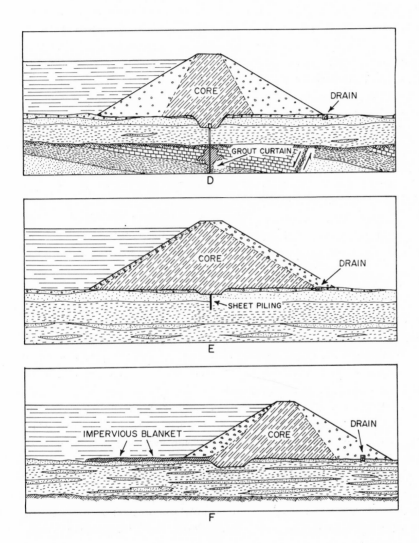

D. Cut-off extends to layer of impervious material in unconsolidated valley fill. Grout holes extend through a limestone layer in bedrock below valley fill.
E. A cut-off is provided by sheet piling driven into an impervious layer in valley fill.
F. Flow beneath dam is reduced by a layer of impervious material placed upstream from the dam.

Grout is a liquid, either a uniform chemical substance or an aqueous suspension of solids that is injected into rocks or unconsolidated materials through specially drilled boreholes to improve bulk physical properties and/or to reduce or eliminate seepage of groundwater. Grouting materials are of three basic types: (1) Portland cement-base slurries, (2) chemical grouting solutions, and (3) organic resins, including epoxy resins (polymers). Portland cement slurries are by far most widely used in grouting and by addition of various substances such as clay, sand, and bentonite or addition of chemicals to increase or reduce setting time, are used in a wide range of applications (Lenahan, 1973). Water-base chemical grouting materials are used primarily where interstitial openings or cracks are so small that they will not permit circulation and penetration by particle suspensions. Commonly two chemical grouting solutions are mixed immediately prior to or during injection so as to set or precipitate dissolved components at the desired location. A common chemical grout solution contains sodium silicate which is converted to a gel by a catalyst dissolved in a second solution.

Except under exceptional circumstances, organic resins are rarely used in dam foundations because of their high cost. A summary of the use of polymer-forming materials in improving the strength of porous rock has been prepared by Crow and Kelsh (1971).

In dam foundations three kinds of grouting programs are identified: (1) comparatively shallow systematic "blanket" or "consolidation" grouting over critical portions of the foundation, (2) "curtain" grouting from a gallery or concrete "grout cap" along a specified location to produce a deep impermeable barrier to subsurface groundwater seepage, and (3) "off-pattern", special purpose grouting to improve strength and/or to overcome problems created by groundwater circulation in zones identified by field geotechnical studies.

Although grouting of a rock foundation may be conducted with meticulous care, the possibility always exists that some channelways of underground water circulation remain and that flow through these channelways will accelerate as the reservoir is filled. If the volumes of flows prove to be excessive during reservoir filling, immediate remedial steps must be taken, but, if the flows are small or insignificant, they may be intercepted and diverted by drain holes or porous prisms. Interception and diversion provide an opportunity for constant visual or instrumental monitoring of seepage beneath a dam as it varies with filling and drawdown of the reservoir behind the dam.

Several hypothetical cross-sections of dams constructed on bedrock showing the locations of foundation grout holes and drainage holes or prisms are illustrated in Fig.10-2. The various kinds of holes are given separate letter designations as follows: "*A*" = curtain grout holes; "*B*" = blanket grout holes; "*C*" = special purpose, off-pattern grout holes; "*D*" = drain holes.

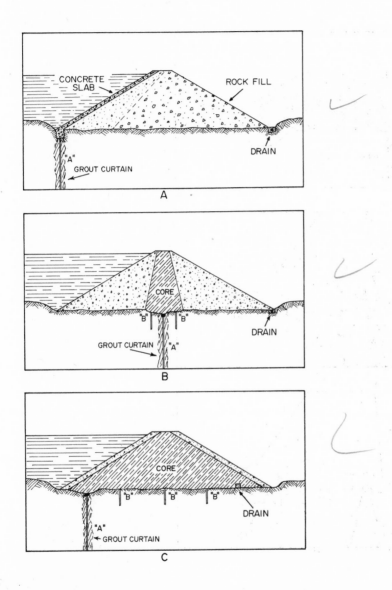

Fig.10-2. Some cross-sections of dams with rock foundations showing locations of drilled holes for foundation treatment. See text discussion for explanation of letter designations for holes. (Continued on p. 240.)
A. Rock-fill dam with impermeable concrete face.
B. Zoned earth and rock-fill dam.
C. Zoned earth and rock-fill dam.

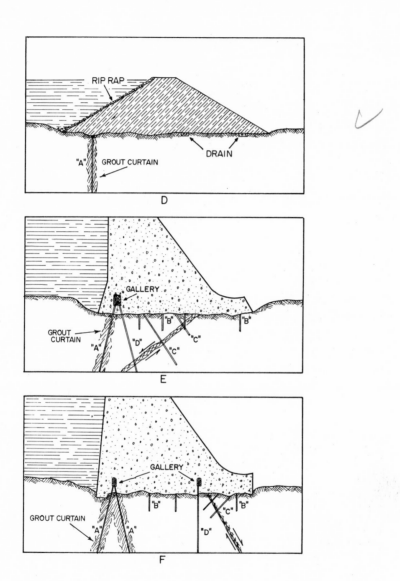

Fig. 10-2 (continued).
D. Earth dam.
E. Concrete gravity dam with "*C*" holes intersecting a fault zone.
F. Concrete gravity dam with double grout curtain and "*C*" holes intersecting a permeable fault zone.

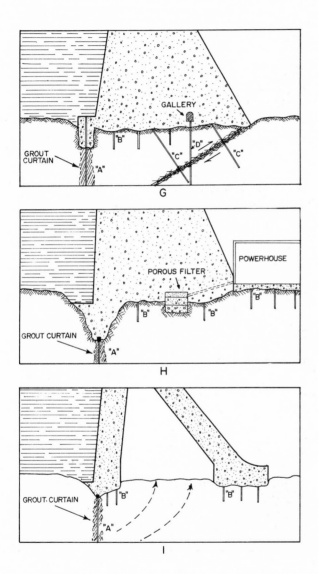

G. Concrete gravity dam with special purpose "*C*" holes.
H. Concrete gravity dam with a porous filter to collect seepage water.
I. Hollow concrete gravity dam. Arrows indicate possible flow of seepage water past grout curtain.

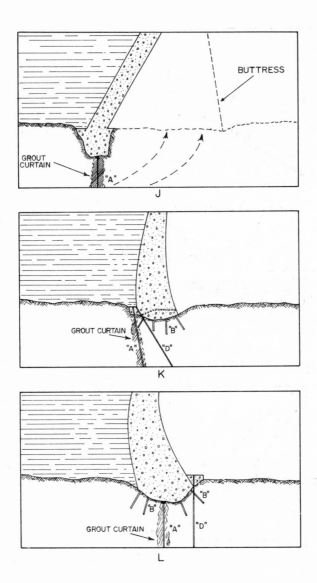

Fig.10-2 (continued).
J. Concrete buttress dam. Arrows indicate possible seepage past grout curtain.
K. Concrete arch dam.
L. Concrete arch dam.

INTERVAL GROUTING WITH PACKERS

In blanket grouting, especially where an exposed foundation enables direct observation of the consequences of grout injection, there is generally no need for localizing the penetration of subsurface materials by use of packers. In contrast, knowledge of the depths of permeable zones in certain grout holes is essential for the efficient control of depths of intermediate holes in closure drilling patterns.

Two methods are in common use in confining grout injection to fixed intervals in curtain and deep, off-pattern grout holes. In one method the grout hole is drilled to its full depth and grouting is accomplished by systematic elevation or lowering in the hole of two packers a fixed distance apart, say 20 ft. This method has the disadvantage that cuttings from the drill bit tend to fill openings in the walls of the core hole in its upper portions and impede entrance of grout suspensions. Further, if difficulty is encountered in drilling the hole, perhaps because of closely fractured rock, special measures, such as cementing, may be required to overcome the difficulty, and may reduce total grout take.

The second method, preferred by the writer, utilizes a single packer which is set at the top of a newly drilled interval of 10–20 ft of hole and in a section of the hole previously grouted by the same procedure. This method permits close estimation of the locations of permeable zones and at the same time enables efficient penetration of intervals of the grout holes that might tend to cave during drilling or that contain groundwater under a pressure head.

PATTERN GROUTING

Plans for dams commonly include broad specifications for a systematic program of blanket and/or curtain grouting. However, because of the uncertainty as to the conditions that will be encountered during the grouting operations, the number and depths of grout holes are not precisely stated. Instead, responsibility for completion of an adequate program is delegated to the field or construction engineer who is instructed to conduct grouting operations "as required" or "as necessary". Except under the most favorable circumstances, exact prediction of the total amount of grout that will be required is extremely difficult. Grout "take" in amounts moderately to greatly in excess of estimates made prior to construction is a common experience, and confirms a widely prevalent statement that "grouting is an art and not a science".

"Pattern" grouting is grouting included in the plans and specifications for a dam and commonly is the basis for estimation prior to construction of the total footage of grout holes and the expected amount of grout consumption. It is general practice to lay out locations of grout holes in the plans with a definite, systematic pattern, spacing, and assumed depths. If sufficient geological information is at hand, the locations and depths of pattern holes take into account the three-dimensional geometry of geologic features.

Locations of pattern grout holes as they might be indicated on plans and specifications for several types of dams are shown in Fig.10-3—10-5. The drawings are entirely schematic, and no scale is shown, on the assumption that the actual number of holes will be determined by the area and cross-sectional configuration of the excavation for the dam foundation. In addition, it is assumed that in each example the dam foundation is entirely in bedrock, so that both blanket and curtain grouting are anticipated in plans and specifications.

BLANKET GROUTING

"Blanket" grout holes ("*B*" holes) usually are shallow, not more than 20—30 ft and are intended to remedy flaws in the foundation, such as fractured rock, by reducing permeability and increasing bulk strength. Although blanket

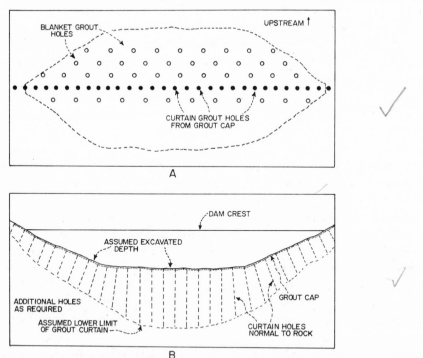

Fig.10-3. Schematic locations of pattern blanket and curtain grout holes in bedrock of an earth dam of moderate size.
A. Plan.
B. Section showing formula depths for curtain grout holes.

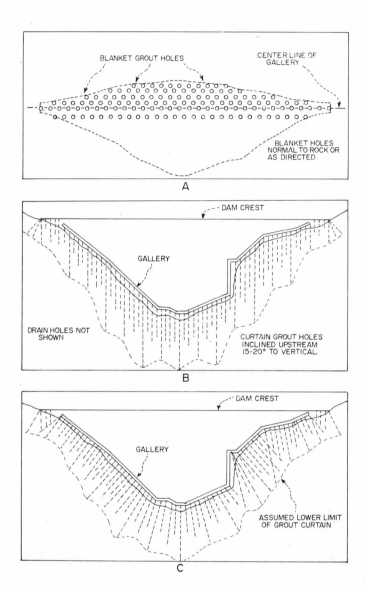

Fig.10-4. Schematic locations of pattern blanket and curtain grout holes in foundation of a concrete gravity dam. Curtain grout holes are drilled from a gallery within the dam.

A. Plan showing locations of blanket holes in bedrock.

B. Section showing locations of curtain grout holes (preferred pattern).

C. Section showing alternate plan for grouting from grout gallery. Note fanning of holes with depth, generally undesirable in deeply fractured bedrock.

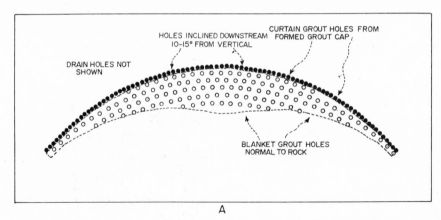

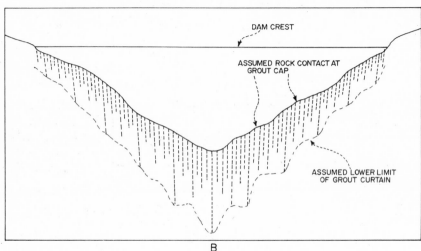

Fig.10-5. Schematic locations of pattern blanket and curtain grout holes in bedrock foundation of a thin-arch dam.
A. Plan.
B. Projection into a vertical section passing through dam abutments. Blanket grout holes not shown.

holes may be routinely drilled normal to the foundation surface, there is considerable merit in directing the holes to intersect specific local features identified in the dam foundation during excavation. "Blanket" grouting must be completed before construction of a dam.

CURTAIN GROUTING

"Curtain" grouting in earth and/or rock-fill dams usually is completed *before* a dam is constructed and is performed from a grout cap consisting of concrete filling a shallow, narrow excavated trench in the foundation. A notable exception to the usual timing of the grouting operation is grouting *after* construction from a grout cap at the upstream heel of a dam. An example of a grout cap in the bottom of a cut-off excavation for a large earth dam is shown in Fig.10-6.

Fig.10-6. Grout cap in bedrock at bottom of cut-off excavation in gravel deposits. Dillon Dam, Colorado. (Photo courtesy of Denver Board of Water Commissioners.)

Curtain grouting of foundations of concrete dams is most effective after completion of the dam, at a time when the full load of the dam is being exerted on the foundation. Under such circumstances higher pressures may be used in grouting so as to assure maximum travel of grout in all directions along flow paths intersected by grout holes.

In gravity and gravity-arch dams of moderate to large size it is common practice to construct a gallery inside the dam for drilling curtain grout holes ("*A*" holes) and drainage holes ("*D*" holes). Foundations of small gravity- and thin-arch dams are efficiently grouted from grout caps along the contact of the upstream face of the dam with rock.

In the absence of geological controls that dictate otherwise, the depths of grout holes in curtain pattern grouting are determined by formula. A commonly used formula states that the vertical depth of a curtain grout hole shall be a third of the height of the dam at the location of the hole plus 50 ft (or 15—20 m). Where geological conditions in the foundation are known, the depths of curtain holes are not based on a formula, but, instead, are determined by the locations at depth of features requiring grout injection to remedy or improve their physical properties and to reduce or eliminate potential seepage of groundwater through them.

The spacing and sequence of pattern drilling and grouting of curtain grout holes, either from a grout cap or a gallery, is of utmost importance. Commonly, planned curtain grout holes are spaced 10 ft (or 3 m) apart, the distance between them measured as a slope distance or a horizontal distance, with implied provisions for additional holes as required. In accepted practice the sequence of drilling and grouting is controlled by a predetermined closure pattern such as the one indicated in Fig.10-7, which shows an 80-ft (or 24 m) closure pattern for bedrock of uniform properties and low average permeability.

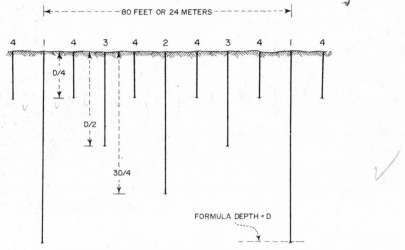

Fig.10-7. Conventional closure pattern for curtain drilling and grouting. Numbers indicate sequence of drilling and grouting.

In Fig.10-7 it is assumed that grout take in each of the holes is insignificant and that, as the pattern is closed, holes are drilled to successively shallower formula depths. In an actual example determination of the depths of intermediate holes is controlled by the experience in preceding holes, thus lending great versatility to the procedure.

Except in situations where foundation materials are uniform and geologic structures are very simple, blanket and curtain grouting operations should be conducted under constant and close supervision by an experienced geologist. The knowledge of foundation conditions obtained during preliminary and design-stage investigations almost invariably does not provide a complete understanding of the exact locations, dimensions, and properties of subsurface geologic features that may

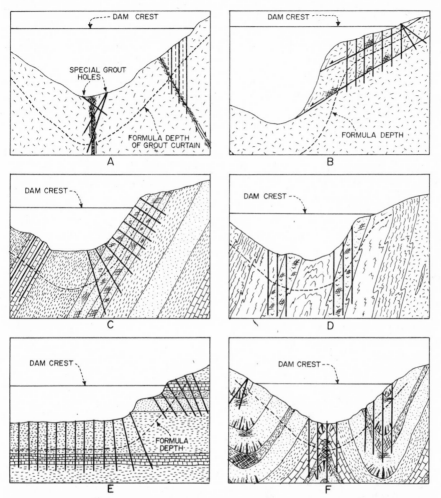

Fig.10-8. Curtain grout holes with depths and attitudes determined by subsurface geologic conditions. Curved dashed lines indicate maximum depths of grout curtain as they might be calculated by a formula without regard to geologic features.

A. Special inclined holes are drilled to intersect a wide broken zone along a steep fault, and pattern grout holes are deepened to intersect another fault in an abutment.

B. Pattern grout holes are deepened to intersect permeable zones along gravity-slip faults in a steep-walled canyon in crystalline rocks.

C. Pattern grout holes are deepened in a limestone layer and are extended to cross an inclined layer of brittle, fractured sandstone.

D. Grout holes are deepened to penetrate fractured layers of brittle quartzite in a metamorphic sequence containing schists and gneisses.

E. Grout holes are extended beyond formula depths to test for solution permeability in limestone.

F. Deep grout holes probe highly-fractured zones associated with tight folds in sedimentary rocks.

promote adjustments in the foundation during loading or allow excessive ground-water seepage.

Examination of cuttings from grout holes and, if deemed necessary, recovery and examination of cores, enables logging of gross- and small-scale features that contribute to a growing appreciation of the details of subsurface geology and its increasingly accurate reconstruction. Water testing under pressure is not required because the measured consumption of grout serves the same purpose, particularly if initial mixtures are thin.

Fig.10-8 suggests the requirement for careful consideration of subsurface geology in locating and determining the depths of curtain grout holes in a variety of situations, so that potentially dangerous groundwater seepage beneath or at the sides of the formula limit for the grout curtain may be reduced or eliminated. Only grout holes of special locations or depths (heavy lines) in excess of those calculated by formula are shown. It is assumed that pattern spacings and formula depths of other holes will be maintained as the grout curtain is completed for the entire foundation.

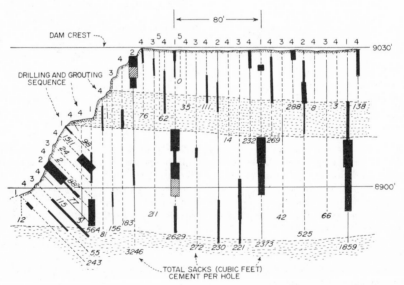

Fig.10-9. Grout curtain along a segment of the foundation and abutment of Dillon Dam, Colorado, an earth dam resting on sedimentary rocks and constructed by the Denver Board of Water Commissioners. Shale layers are indicated by appropriate symbols. Rocks in layers above and below shale layers consist mainly of brittle sandstones and sedimentary quartzites (see Fig.10-10). Widths of bars are proportional to grout take between packer settings.

EXAMPLES OF CURTAIN GROUTING

Fig.10-9—10-12 illustrate examples of curtain grouting in which depths, grouting pressures, and grout consistencies were controlled by the foundation

Fig.10-10. Closely fractured, highly permeable quartzite exposed in grout trench during construction of Dillon Dam, Colorado. (Photo courtesy of Denver Board of Water Commissioners.)

geology. In each example an understanding of the foundation geology was obtained from exploratory core holes drilled prior to design and construction of the dam, detailed surface geologic mapping prior to and during dam construction, and continued observation of the progress of the grouting operation. At each dam an 80-foot closure pattern was employed, and, as required, additional grout holes were drilled to remedy special conditions.

Fig.10-9 illustrates the results of curtain grouting of a segment of the foundation of Dillon Dam, Colorado, an earth dam situated on a succession of faulted, tilted, and, locally, closely fractured sedimentary rocks. A particular problem was presented by close fracturing in brittle sandstones and quartzites in the Dakota formation (Fig.10-10), which, prior to grouting, were highly permeable to groundwater circulation.

Fig.10-11 is a projection on a vertical plane of the grout curtain at the Williams Fork Dam, Colorado, a thin-wall concrete arch dam. Grout holes were drilled from a grout cap at the upstream contact of the dam with rock at angles from vertical to 60° from the vertical and inclined under the dam. Preliminary geological investigations and subsequent examination of the dam foundation during excavation revealed the existence of a system of essentially vertical faults of small displacement and intersecting crystalline metamorphic gneisses and schists. The trend of the faults is generally across the axis of the dam, and their existence created the probability of extensive seepage beneath the dam unless they were

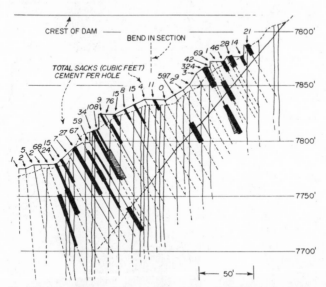

Fig.10-11. A portion of the grout curtain at the Williams Fork Dam, Colorado, a thin-arch dam resting on closely sheared crystalline rocks. Widths of bars are proportional to grout take between packer settings. Projection on a vertical plane. Looking downstream. Dam constructed by Denver Board of Water Commissioners.

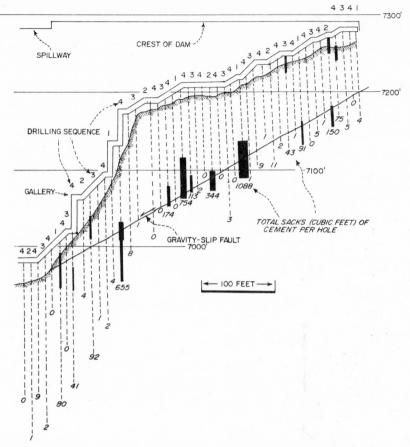

Fig.10-12. A portion of the grout curtain at Gross Dam, Colorado. Note extensive penetration of a gravity-slip fault by grout. Dam constructed by Denver Board of Water Commissioners.

sealed with grout. During the grouting operation a gravity-slip fault, not previously known to exist, was discovered and treated.

Fig.10-12 shows a portion of the grout curtain drilled from a gallery in Gross Dam, Colorado, a concrete gravity-arch dam resting on massive granite. Grout penetration was especially notable along a gravity-slip fault dipping toward the canyon floor.

OFF-PATTERN, SPECIAL-PURPOSE GROUTING

During investigations prior to dam construction or as unanticipated geological conditions are exposed in foundation excavations, the need for "off-pattern"

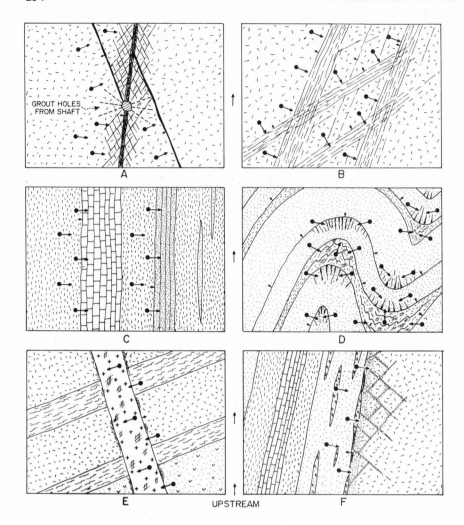

Fig.10-13. Off-pattern, special-purpose grout holes ("C" holes) in bedrock foundations.
A. Inclined holes from the surface and horizontal holes from a shaft intersect steep faults and associated fractures at depth.
B. Inclined holes are drilled to intersect downward projections of sheeted zones in brittle crystalline rocks.
C. Tilted sedimentary layers present a potential for seepage under a dam. Inclined holes are drilled to intersect a limestone layer and a brittle sandstone layer.
D. Inclined, off-pattern holes are drilled into jointed and sheared rock in crests and troughs of folds in sedimentary rocks.
E. Holes are drilled to intersect a closely jointed igneous dike at depth.
F. Off-pattern grout holes intersect a jointed, weathered zone in crystalline rocks below an unconformity.

special-purpose grout holes ("C" holes) may become apparent. These holes are drilled and grouted to improve the strength and/or reduce the permeability of rock masses that are not intersected by blanket or curtain grout holes.

Examples of foundation conditions requiring off-pattern grouting are shown in Fig.10-13. The depths, directions, and inclinations of the grout holes are determined by the three-dimensional geometry of zones of incompetent and/or permeable rocks as revealed by field examination of bedrock exposures in foundation and abutment excavations.

GROUT CONSISTENCY AND GROUTING PRESSURES

The purpose of foundation grouting is improvement of the bulk strength of foundation materials, and, generally an even more important consideration, filling and sealing off of potential avenues of groundwater seepage. Accomplishment of this purpose under many circumstances requires the exercise of great skill and judgment in location of grout holes, determination of depths, control of grouting pressures, and control of grout consistencies.

Locations and depths of grout holes are determined by a knowledge of local geological conditions and by the experience obtained during progress of a carefully supervised grouting operation, but pressures and consistencies of grout must be based on highly subjective considerations. Accordingly, it is not surprising that there are many widely divergent opinions and prescribed procedures for surface control of grouting operations.

The ability of cement grout to penetrate interconnected open spaces is limited by the dimensions of the open spaces and the amount and size of the cement particles suspended in the water base. Openings of slightly greater than capillary size that may permit free circulation of groundwater are quickly filled and obstructed by cement particles and lateral and/or vertical travel of the grout suspension is greatly impeded or brought to a halt. In larger openings, presupposing interconnecting avenues of circulation, grout suspensions move with ease and in some instances travel surprisingly large distances.

If easy grout circulation continues with the progress of the grouting operation, the suspension is gradually thickened and, if necessary, the pressure correspondingly increased until filling of available openings is indicated by refusal of the grout hole to accept additional grout. Grout leaks at the surface should be calked or otherwise sealed to promote confined subsurface movement of grout suspensions.

The definitions of "thin" and "thick" are not precise, but, generally, "thin" mixtures are construed to mean mixtures prepared by mixing 8—10 volumes of water with one volume of cement. "Thick" mixtures have volume proportions of cement to water of approximately 1:1, or thicknesses that are not so great that the grout can not be pumped with reasonable ease. In practice, experimentation with proportions of cement to water in the initial stages of grouting to determine optimum conditions for circulation often serves a useful purpose.

Under some circumstances, especially in extensively fractured rocks, in cavernous soluble rocks such as limestones, and in highly permeable gravels, initially thick grout mixtures are indicated, and inert additives such as clay or sand may be added to grout suspensions as inexpensive fillers.

Determination of the pressures used in grouting operations requires exercise of considerable judgment, and depends on the nature of local conditions. When grout pressures exceed certain critical limits, the possibility exists that foundation rocks may be dislocated, and channelways of circulation that did not exist previously may be newly created. The possible consequences of use of excessive grout pressures are understood by reference to Fig.10-14 in which it is supposed (Fig.10-14A) that a grout hole intersects a horizontal channelway of potential grout circulation, such as a flat fracture or a sedimentary bedding plane and that

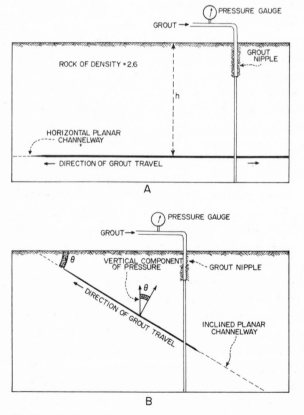

Fig.10-14. Horizontal and inclined planar channelways of grout circulation. See text discussion.

(Fig.10-14B) a grout hole intersects a similar, but inclined, plane of potential circulation.

In Fig.10-14A the depth of a horizontal confined channelway is indicated as *h,* and it is assumed that the pressure of the grout suspension in the grout hole at the point of entrance into the channelway is very nearly the same as the pressure indicated by a pressure gauge at the surface. Entrance of grout into the channelway dislocates the rock mass above the channelway, and by wedge action and local elastic and/or plastic adjustments in the rock tends to widen the channel so as to permit lateral spread of the grout suspension. As for movement of groundwater, friction along the channelway reduces the fluid pressure at any point as a function of distance from the grout hole, but as the channelway widens, friction becomes of decreasing importance so that grout pressures in the channelway begin to approach the same order of magnitude as pressures recorded at the pressure gauge.

The lifting force of grout under pressure in a horizontal channelway as a function of pressure at any point in the channelway can be easily calculated. Assuming a rock density of 2.6, near that of average granite, Table 10-1 was

TABLE 10-1

Height of column of rock of density 2.6 that can be displaced upward by grout under pressure in a confined horizontal channelway

Pressure (psi)	h (ft)
10	8.8
50	44.4
100	88.8
200	177.7
300	266.6
400	355.5
500	444.4
1000	888.8

prepared and indicates the height, *h,* of a column of rock that can be lifted as a function of the pressure at any point along the horizontal channelway. The table is calculated by determining that the vertical pressure exerted by a column of rock of density 2.6 increases by 1.125 psi per added foot of height.

In Fig.10-14B a planar channelway inclined at an angle θ to the horizontal is shown. The pressure exerted vertically upward by the grout at any point in the channelway is obtained by multiplying the grout pressure by $\cos\theta$. Clearly the weight of the overlying column of rock at any point decreases as the grout approaches the surface, thus explaining the increasingly easy movement of the grout toward the surface. When $\theta = 90°$, that is, for vertical channelways of circulation, the entire pressure of the grout is directed horizontally, and high pressures are unlikely to cause extensive dislocations of rocks, except in steep valley slopes.

Because of a wide range in complexity of patterns of underground circulation it is not possible to establish a rigid formula for controlling grout pressures at the top of a grout hole. For reasons outlined above, care should be taken in maintaining low grout pressures where grout holes intersect channelways in the floors of valleys and channelways paralleling the sides of valleys, especially during blanket grouting. For curtain grouting, a rule that is sometimes followed states that pressure in an initially thin grout suspension is increased to a level which establishes a free circulation (assuming that channelways for circulation are present) but not in excess of the calculated hydrostatic pressure of the filled reservoir at the elevation of the collar of the grout hole plus 10—50 psi. The hydrostatic pressure of water increases 0.433 psi per foot of depth. Of course, the hydrostatic pressure of the grout in the grout holes increases with depth, but this pressure is usually ignored.

As grout mixtures are thickened, and as a grout hole approaches grout refusal, pressures may be gradually increased, but except in deep holes in high-strength rocks, should not be allowed to exceed a pressure of twice the calculated hydrostatic head of the filled reservoir at the elevation of the collar of the grout hole plus 10—50 psi.

A not uncommon experience is a quick drop in pressure at the gauge when grout suddenly forces its way into a new channelway of easier circulation. When this happens, grout pressures are reduced and continued at a level just sufficient to maintain circulation.

Premature thickening of grout or reduction of pressures to cause grout refusal in a grout hole should be avoided unless it can be demonstrated that grout is escaping to the surface well outside of the foundation area. So long as grout is circulating somewhere in the foundation of a dam or in the near proximity of the foundation, it must be assumed that it is contributing to an improvement of the engineering properties of foundation materials and to a reduction in permeability to groundwater seepage.

DRAINAGE AND OBSERVATION HOLES, WELLS, AND POROUS PRISMS

In spite of carefully executed curtain and blanket grouting operations the possibility, and indeed the probability, exists that all possible avenues of circulation of groundwater have not been intersected and sealed by grout. Assessment of the effectiveness of the grouting operation usually is not possible until the reservoir behind the dam is partly to completely filled. Accordingly, it is standard practice to drill holes, excavate wells, or construct porous drainage prisms downstream from grout curtains to intercept groundwater that passes through or beneath the curtains and to enable observation of changing volumes of flow with filling and drawdown of the reservoir.

In an earth dam which buries the grout cap excessive seepage through the curtain after reservoir filling presents a major problem that can not be remedied by any simple means. To prevent such a happening it is apparent that both blanket and curtain grouting prior to dam emplacement must be conducted with great care and meticulous attention to details of foundation geology.

In concrete gravity and gravity-arch dams which contain grout galleries and in thin-arch dams remedial grouting utilizing existing drain holes or newly drilled holes presents no insurmountable problems.

ROCK REINFORCEMENT BY ROCK BOLTS AND STEEL CABLES

Fractured rock masses or weak sedimentary rocks in which planes of weakness are inclined toward a valley floor, and closely and complexly fractured masses, without regard to the orientations of the fractures, tend to be unstable and may present a constant threat of collapse by slope failure, especially if their properties are changed by exposure to the atmosphere or the masses are penetrated by water under pressure from a reservoir. Within dam excavations fractured and layered rocks are contained by the dam, and slope failure can no longer be considered a problem. Outside of and above a dam, however, unstable masses may become activated so as to cause partial filling of a reservoir, obstruction of diversion structures, or physical damage to the dam and appurtenant features.

Although confinement by the dam of closely fractured masses within a foundation excavation prevents slope failures, such masses may have inadequate bearing strength and, if the fractures are oriented in critical directions, may contribute to uneven dislocation and failure of a dam.

Grout treatment of fractured foundation materials increases bulk strength and reduces permeabilities, but in some instances reinforcement by rock bolts and/or tensioned steel cables is required to stabilize and provide the additional strength to provide a necessary adequate factor of safety against failure either in the dam foundation or in its vicinity.

Rock bolts and/or steel cables, if properly utilized, increase the strength of rock masses by tending to close open fissures and by increasing shearing strengths along fractures and weak layers by increasing frictional resistance.

Rock bolts and steel cables are not particularly effective unless they are firmly anchored so that they may be tensioned. Rock bolts commonly have a short expanding anchoring device at the bottom end and require firm rock for effective anchoring. Increasing use is being made of quick-set, high-strength resins to anchor both rock bolts and steel cables. Resins perform well even in materials of low to moderate strength and have the advantage of providing an anchored interval of any desired length by controlling the volume of resin poured into a drilled hole. Cables are used in holes deeper or longer than those normally drilled for rock bolts, but may be used instead of rock bolts under many circumstances.

Proper use of rock bolts or steel cables requires an understanding of frictional properties of planes of weakness and the shearing stresses that cause dislocations along these planes.

The *coefficient of static friction, K,* (Fig.10-15) by definition is the ratio of

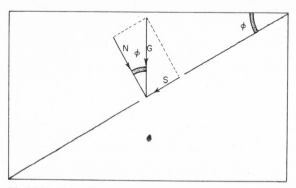

Fig.10-15. Force diagram determining coefficient of static friction.

the shear force, *S,* required to initiate movement of one surface over another to the force, *N,* acting normal to the contact between the surfaces and tending to press them together. If we convert force to stress per unit area and taking into account the angular relations in Fig.10-15, we have:

$$K = S/N = \sigma_S/\sigma_n = \tan \phi \qquad\qquad (10\text{-}1)$$

In Fig.10-15 *S* and *N* are considered to be vectorial components of *G,* the force of gravity.

After initiation of movement along the contact surface under a driving force parallel to *S,* the frictional relationships are specified by a *coefficient of kinetic friction,* which generally has a smaller value than the coefficient of static friction.

In the eighteenth century Coulomb, a French physicist, observed that the shearing strength (shearing resistance) associated with sliding of one surface past another depends on two factors which he called *internal friction* and *cohesion,* respectively, and established the empirical relationship:

$$\qquad\qquad (10\text{-}2)$$

The value for *c* depends on the smoothness or roughness of the surfaces in contact and, in some situations, on the presence or absence of pore water or an infilling of a weak material such as clay. Rough surfaces of contact in brittle substances promote *keying*, which resists dislocations until they are removed by attrition or crushing. In many rocks movement along fractures is accompanied by a reduction in the value of c so that it approaches zero. In engineering calculations designed to insure an adequate factor of safety, it is common practice to disregard the value of c, thus giving values for the shearing strength less than the actual strength.

Water under pressure in fractures or in pores acts to reduce the effective normal stress by an amount equal to the pressure. In Fig.10-15 positive piezometric pressure opposes the force vector, N, and simultaneously reduces the effective values of G and S thus reducing the shearing strength along the surface. This condition is expressed by:

$$S = (\sigma_n - u)\tan \phi + c \tag{10-3}$$

in which u is the piezometric pressure.

Sliding of one surface past another can be induced by increasing the force S acting in the direction of surface of potential dislocation or by increasing the angle ϕ of the surface (Fig.10-15). That angle of frictional resistance at which spontaneous sliding can occur, ϕ_{crit}, is sometimes called the *critical angle of repose* or more simply, the *angle of repose* and can be calculated from eq. 10-3 when S, σ_n, u, and c have been determined experimentally either for dislocations along a fracture or along a plane of rock weakness such as a bedding plane or foliation plane. Calculated angles of repose in rocks commonly are of the order of $35-50°$, except in weak rocks, including poorly indurated shales and claystones, which may have angles of repose as low as about $20°$. In very weak unconsolidated water-bearing materials the value for ϕ_{crit} commonly lies in the range $10-20°$.

In natural slopes in valleys in which fractures or layered weak materials dip toward the valley floor the angles of repose rarely exceed $40°$. The highest angles of repose are observed in strong rocks in which surfaces of contact along planes of fracture are rough and well keyed. A useful procedure is field measurement of dip angles of planar structures which, where exceeded, have resulted in slope failure.

Some critical angles of repose and the corresponding coefficients of static friction are given in Table 10-2.

Reinforcement of rocks in slopes within and in the vicinity of dam excavations is designed to stabilize and improve the strength of masses in which fractures or layering is inclined toward the valley floor or to stabilize complexly fractured masses which might collapse by mass movement, either spontaneously or because of oversteepening of slopes or removal of support during excavation. Processes which promote sliding along planes of fracture or weakness, in effect reducing the critical angle of repose, include penetration by water under pressure, wetting and expan-

TABLE 10-2

Critical angles of repose and corresponding coefficients of static friction

Angle of repose (ϕ_{crit}) (degrees)	$K = \tan \phi$
10	0.18
20	0.36
30	0.58
40	0.84
50	1.19

sion of clay in filling, frost action, and, where the planar structure is almost at the angle of repose, sudden dislocations by earth tremors.

Where rock bolts or anchored steel cables are used for reinforcement, they should be installed as nearly at right angles to planes of fracture or weakness as possible. The reason for this becomes apparent from inspection of Fig.10-16 which

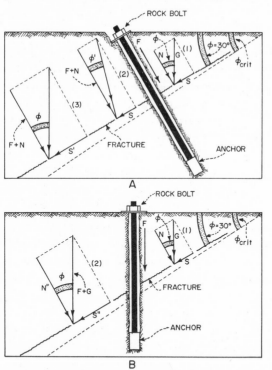

Fig.10-16. Rock bolts intersecting a fracture which may localize shear dislocation. See text discussion.
A. Rock bolt is normal to fracture.
B. Rock bolt is vertical and creates a force parallel to the force of gravity.

shows a fracture inclined to the surface at an angle of 30°. The angle ϕ is slightly less than the critical angle, ϕ_{crit}, which is the angle of repose, but to be on the safe side in calculations, ϕ is assumed to be equal to ϕ_{crit}. Actually, there has been no sliding dislocation along the fracture because $\phi < \phi_{crit}$ by a small but unknown amount.

In Fig.10-16A a force diagram, (1), gives a coefficient of static friction, K, equal to $S/N = \tan 30^{\circ} = 0.577$, assuming that the angle of repose is 30°. Resolution of the force of gravity, G, gives a component, N, normal to the fracture and a component, S, which is the force tending to cause sliding dislocation along the fracture. The effects of cohesion, c, in eq.10-2 are not considered. In Fig.10-16A a tensioned rock bolt produces a force, F, acting in the same direction as N, and arbitrarily set equal to N. Because S remains constant and the force normal to the fracture is now $F + N = 2N$, a new force diagram (2) results, and the value for $\tan \phi'$ $= S/(F+N) = 0.27$ which corresponds to the coefficient of friction and an angle of repose of an hypothetical material much weaker than the actual material under consideration.

To determine the force, S', necessary to initiate sliding dislocation along the fracture after installation and tensioning of the rock bolt another force diagram, (3), is constructed. From the relationship $K = S/N = S'/(F+N) = \tan 30^{\circ} = 0.577$ it is determined that $S' = 2S$, that is, installation and tensioning of the rock bolt has *doubled* the shearing resistance along the fracture. In addition, the rock bolt itself resists shear dislocation.

In Fig.10-16B, the rock bolt is installed vertically, parallel to the direction of the force of gravity, G. The force diagram before installation of the bolt, (1), is identical with the initial force diagram in Fig.10-16A. In force diagram (2) it is arbitrarily assumed that the force exerted by the rock bolt, F, is equal to the force of gravity, G, and a new shearing force, S'', is calculated. However, although S'' is twice S, the ratio S''/N'' in force diagram (2) is the same as S/N in diagram (1), and both ratios equal K, the coefficient of sliding friction. Thus, the only contribution that the rock bolt makes toward preventing sliding is the strength of the bolt itself except, possibly, for an increase in the value of c, the factor for cohesion in eq.10-2 or 10-3.

It is usually not possible to calculate spacings and tensioning for rock bolts (or steel cables) required to stabilize and provide a predetermined factor of safety in complexly fractured rocks. However, at times the space geometry enables simple calculations that prove to be very useful. An example is provided in Fig.10-17, in which it is assumed that a rock slab is resting nearly at the angle of repose on a fracture dipping 35° toward the valley floor and striking parallel to the valley. For purposes of calculation it is assumed that the slab extends for 50 ft parallel to the valley floor and the other dimensions are those shown in the diagram. Pertinent data are as follows:

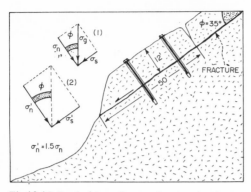

Fig. 10-17. Rock slab resting nearly at angle of repose on a fracture. See text discussion.

Volume of slab = 27,000 cub. ft
Weight (160 lbs/cub. ft) = 4,320,000 lbs
Area of fracture beneath slab = 2500 sq. ft = 360,000/sq. inch.
$K = \tan \phi = \tan 35° = 0.700$

Calculations based on the stress diagram, (1), in Fig. 10-17 yield the following numbers:

$\sigma_g = 12$ psi
$\sigma_n = \cos 35° = 9.83$ psi
$\sigma_s = \sigma_n \sin 35° = 6.87$ psi and $\sigma_s/\sigma_n = \tan \phi$

Now, suppose that it is desired to increase σ_s by 50% by installation of rock bolts perpendicular to the fracture and anchored in firm rock. That is, it is wished to determine a value for σ_n', stress diagram (2), which corresponds to $\sigma_s' = 1.5 \ \sigma_s = 10.31$ psi for an angle of repose of 35°. Accordingly, we now calculate a value for σ_n' using the relationship $\sigma_n'/\sigma_s' = \tan \phi$, and conclude that the rock bolts must supply an additional stress normal to the fracture equal to 4.91 psi to realize the desired objective. Thus:

Desired additional normal stress = 4.91 psi
Area of fracture = 360,000 sq. inch
Desired additional normal pressure on fracture = 360,000 × 4.91 = 1,767,600 lbs

Suppose that the rock bolts that are installed will be tensioned to 10,000 lbs. Then to obtain the required number of rock bolts we divide 1,767,600 by 10,000 and arrive at 176 rock bolts, that is, approximately one rock bolt per 14 sq. ft. Further increasing the shear strength along the fracture is the strength of the bolts which must be sheared before movement may be initiated.

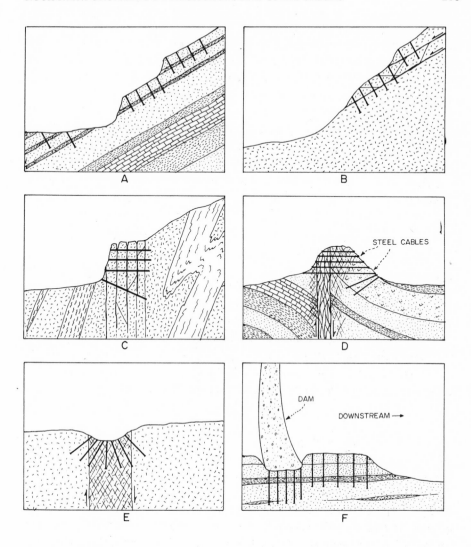

Fig. 10-18. Reinforcement of rock masses by tensioned rock bolts or steel cables.

A. A potentially unstable slope in inclined sedimentary rocks is stabilized with rock bolts.

B. Blocks of crystalline rocks above faults inclined toward the valley floor are pinned by tensioned rock bolts.

C. Rock bolts are used to add strength to a crystalline rock mass intersected by vertical shears. Grouting after tensioning of the rock bolts will serve a useful purpose.

D. A shattered zone along a fault zone is stabilized by steel cables and grouting before and after tensioning of the cables.

E. Rock bolts are installed in a closely fractured zone in a foundation. Grouting before and after rock bolt installation is recommended.

F. Rock bolts or heavy steel cables are installed to reduce the hazard of possible dislocation along a horizontal shale layer when reservoir is filled.

In many situations installation of rock bolts requires careful subjective evaluation of the existing conditions. Commonly unknown are the angle of repose which determines the coefficient of friction, the exact geometry and resolution of forces in complexly fractured masses, the extent to which pore water under pressure may modify σ_n, and the value for c in eq.10-3.

Grouting combined with rock bolt or steel-cable installation produces excellent results. In unstable slopes grouting is done after installation of rock bolts. In confined highly fractured masses in foundations grouting should *precede* and *follow* installation of tensioned bolts or cables so as to obtain optimum results.

Several examples of the use of rock bolts and steel cables are shown in Fig.10-18.

RESERVOIR SLOPE STABILIZATION

Where the possibility of massive failure in slopes of reservoirs exists, whether empty or full, appropriate steps should be taken to stabilize the slopes. The particular remedy that is employed depends on the nature of local conditions. Excavation or anchoring of unstable accumulations of unconsolidated materials, and excavation or stabilization by grouting, rock bolts, or steel cables of unstable rock slopes should be considered as routine in situations where the short-term or long-term integrity of the reservoir is threatened by slope failure.

FINAL CONSTRUCTION REPORT

In addition to a summary of all engineering operations and "as-built" drawings, the final construction report should contain meticulously prepared geologic maps and sections of the foundation and abutments of a dam and appurtenant features and tabular and narrative summaries of all aspects of construction related to the treatment of foundation materials to improve their strength or to reduce subsurface seepage.

REFERENCES

Crow, L. J. and Kelsh, D. J., 1973. Can chemical stabilization improve ground support? *Eng. Min. J.*, August Issue, pp.75–77.
Lenahan, T., 1973. There is a place for grouting in underground storage caverns. *Bull. Assoc. Eng. Geologists*, 10: 137–144.
Wahlstrom, E. E., 1973. *Tunneling in Rock*. Elsevier, Amsterdam, 250pp.

INDEX